SECOND EDITION

Learning Android

Marko Gargenta and Masumi Nakamura

Beijing · Cambridge · Farnham · Köln · Sebastopol · Tokyo

Learning Android, Second Edition

by Marko Gargenta and Masumi Nakamura

Printed in the United States of America.

Published by O'Reilly Media, Inc., 1005 Gravenstein Highway North, Sebastopol, CA 95472.

O'Reilly books may be purchased for educational, business, or sales promotional use. Online editions are also available for most titles (*http://my.safaribooksonline.com*). For more information, contact our corporate/institutional sales department: 800-998-9938 or *corporate@oreilly.com*.

Editors: Andy Oram and Rachel Roumeliotis	**Indexer:** Meghan Jones
Production Editor: Kara Ebrahim	**Cover Designer:** Randy Comer
Copyeditor: Kim Cofer	**Interior Designer:** David Futato
Proofreader: Amanda Kersey	**Illustrator:** Rebecca Demarest

January 2014:   Second Edition

Revision History for the Second Edition:

2014-01-08:   First release

See *http://oreilly.com/catalog/errata.csp?isbn=9781449319236* for release details.

ISBN: 978-1-449-31923-6

[LSI]

Table of Contents

Preface

This book sprang from years of delivering the Marakana Android Bootcamp training class to thousands of software developers at some of the largest mobile companies located on four continents around the world. Teaching this class, over time I saw what works and what doesn't. This book is a distilled version of the Android Bootcamp training course that I developed at Marakana and fine-tuned over numerous engagements.

My background is in Java from back before it was even called that. From the beginning, I was very interested in embedded development as a way to program various devices that surround us in everyday life. Because Java primarily took off in web application development, most of my experience in the previous decade has been in building large enterprise systems. Then Android arrived, and once again I became very excited about building software for nontraditional computers. My current interests lie in using Android on devices that may not even resemble a typical phone. Masumi, my coauthor on this updated second edition, brings with him a ton of experience in mobile, in addition to Java.

This book teaches anyone who knows Java (or a similar language) how to develop a reasonably complex Android application. I hope you find this book fairly comprehensive and that you find the example-based learning reasonably motivating. The goal of *Learning Android* is to get you to *think* in Android terms.

What's Inside

Chapter 1, Android Overview
 An introduction to Android and its history.

Chapter 2, Java Review
 Offers a quick review of Java.

Chapter 3, The Stack
   An overview of the Android operating system and all its parts from a very high level.

Chapter 4, Installing and Beginning Use of Android Tools
   Helps you set up your environment for Android application development.

Chapter 5, Main Building Blocks
   Explains the Android components application developers use to put together an app.

Chapter 6, Yamba Project Overview
   Explains the Yamba application that we'll build together throughout this book and use as an example to learn Android's various features.

Chapter 7, Android User Interface
   Explains how to build the user interface for your application.

Chapter 8, Fragments
   Covers the Fragments API, which helps you separate screens within an application.

Chapter 9, Intents, Action Bar, and More
   Covers some of the operating system features that make an application developer's life easier.

Chapter 10, Services
   Covers building an Android service to process background tasks.

Chapter 11, Content Providers
   Explains the Android framework's support for the built-in SQLite database and how to use it to persist the data in your own application.

Chapter 12, Lists and Adapters
   Covers an important feature of Android that allows large datasets to be linked efficiently to relatively small screens.

Chapter 13, Broadcast Receivers
   Explains how to use the publish-subscribe mechanism in Android to respond to various system and user-defined messages.

Chapter 14, App Widgets
   Shows how to design a content provider to share data between applications, in this case using it to enable our app widget to display data on the home screen.

Chapter 15, Networking and Web Overview
   Covers networking.

Chapter 16, Interaction and Animation: Live Wallpaper and Handlers
    Provides a taste of more advanced subjects.

Conventions Used in This Book

The following typographical conventions are used in this book:

Italic
    Indicates new terms, URLs, email addresses, filenames, and file extensions.

`Constant width`
    Used for program listings, as well as within paragraphs to refer to program elements such as variable or function names, databases, data types, environment variables, statements, and keywords.

`Constant width bold`
    Shows commands or other text that should be typed literally by the user.

`Constant width italic`
    Shows text that should be replaced with user-supplied values or by values determined by context.

 This element signifies a tip or suggestion.

 This element signifies a general note.

 This element indicates a warning or caution.

Using Code Examples

Supplemental material (code examples, exercises, etc.) is available for download at *https://github.com/marakana/LearningAndroidYamba*.

This book is here to help you get your job done. In general, if example code is offered with this book, you may use it in your programs and documentation. You do not need to contact us for permission unless you're reproducing a significant portion of the code. For example, writing a program that uses several chunks of code from this book does not require permission. Selling or distributing a CD-ROM of examples from O'Reilly books does require permission. Answering a question by citing this book and quoting example code does not require permission. Incorporating a significant amount of example code from this book into your product's documentation does require permission.

We appreciate, but do not require, attribution. An attribution usually includes the title, author, publisher, and ISBN. For example: "*Learning Android, Second Edition* by Marko Gargenta and Masumi Nakamura (O'Reilly). Copyright 2014 Marko Gargenta and Masumi Nakamura, 978-1-449-31923-6."

If you feel your use of code examples falls outside fair use or the permission given above, feel free to contact us at *permissions@oreilly.com*.

Safari® Books Online

 *Safari Books Online* is an on-demand digital library that delivers expert content in both book and video form from the world's leading authors in technology and business.

Technology professionals, software developers, web designers, and business and creative professionals use Safari Books Online as their primary resource for research, problem solving, learning, and certification training.

Safari Books Online offers a range of product mixes and pricing programs for organizations, government agencies, and individuals. Subscribers have access to thousands of books, training videos, and prepublication manuscripts in one fully searchable database from publishers like O'Reilly Media, Prentice Hall Professional, Addison-Wesley Professional, Microsoft Press, Sams, Que, Peachpit Press, Focal Press, Cisco Press, John Wiley & Sons, Syngress, Morgan Kaufmann, IBM Redbooks, Packt, Adobe Press, FT Press, Apress, Manning, New Riders, McGraw-Hill, Jones & Bartlett, Course Technology, and dozens more. For more information about Safari Books Online, please visit us online.

How to Contact Us

Please address comments and questions concerning this book to the publisher:

O'Reilly Media, Inc.
1005 Gravenstein Highway North
Sebastopol, CA 95472

800-998-9938 (in the United States or Canada)
707-829-0515 (international or local)
707-829-0104 (fax)

We have a web page for this book, where we list errata, examples, and any additional information. You can access this page at *http://oreil.ly/learning-android-2e.*

To comment or ask technical questions about this book, send email to *bookques tions@oreilly.com.*

For more information about our books, courses, conferences, and news, see our website at *http://www.oreilly.com.* Find us on Facebook: *http://facebook.com/oreilly*

Follow us on Twitter: *http://twitter.com/oreillymedia*

Watch us on YouTube: *http://www.youtube.com/oreillymedia*

Acknowledgments

Marko Gargenta

This book is truly a result of outstanding teamwork. First, I'd like to thank my coauthor Masumi and the editor at O'Reilly, Andy Oram. Mas, I know it took longer than we anticipated. Andy, your comments were spot-on and constructive. Thank you for sticking with the project.

I'd like to thank my team at Marakana, now part of Twitter: Aleksandar (Saša) Gargenta, Ken Jones, Blake Meike—for bringing back firsthand feedback from teaching Android both to Marakana clients and at Twitter to Twitter Engineers. This really made the difference in the direction of the book.

And finally, a huge thanks to my wife Lisa, daughter Kylie, and son Kenzo. You guys are the real inspiration for all this work. I love you!

Masumi Nakamura

I would like to thank first and foremost my coauthor Marko for agreeing to a collaboration on this edition—it has been an awesome ride. Also the people over at O'Reilly—Andy Oram, Allyson MacDonald, and Rachel Roumeliotis, who have been patient and wonderful to work with. Also, Blake Meike and Bill Schrickel for their technical comments and corrections, which have been invaluable.

Thanks also goes out to my family—Shinji, Yuri, Jiro, Toshihisa—who have been very encouraging and supportive (even trying out some of the examples that I have written over the years).

Of note is Jessamyn Hodge, who once again put up with me and supported me throughout the process. Thank you from the bottom of my heart.

Finally, I also would like to thank the Android community and Google, without which this book's very topic would not exist.

## CHAPTER 1

Android Overview

In this chapter, you will learn how Android came about. We'll take a look at its history to help us understand its future. As this mobile environment enters a make-or-break year, we look at the key players in this ecosystem, what motivates them, and what strengths and weaknesses they bring to the table.

By the end of this chapter, you will better understand the ecosystem from a business point of view, which should help clarify the technology choices and how they relate to long-term advantages for various platforms.

Android Overview

Android is a comprehensive open source platform designed for mobile devices. It is championed by Google and owned by Open Handset Alliance (*http://www.openhand setalliance.com/*). The goal of the alliance is to "accelerate innovation in mobile and offer consumers a richer, less expensive, and better mobile experience." Android is the vehicle to do so.

As such, Android is revolutionizing the mobile space. For the first time, it is a truly open platform that separates the hardware from the software that runs on it. This allows for a much larger number of devices to run the same applications and creates a much richer ecosystem for developers and consumers.

Let's break down some of these buzzwords and see what's behind them.

Comprehensive

Android is a comprehensive platform, which means it is a complete software stack for a mobile device.

For developers, Android provides all the tools and frameworks for developing mobile apps quickly and easily. The Android SDK is all you need to start developing for

Android; you don't even need a physical device. Yet, there are numerous tools, such as Eclipse, for example, that help make the development life cycle more enjoyable.

For users, Android just works right out of the box. Additionally, users can customize their phone experience substantially. It is, according to some studies, the most desirable mobile operating system in the United States at the moment.

For manufacturers, it is the complete solution for running their devices. Other than some hardware-specific drivers, Android provides everything else to make their devices work. That means that manufacturers can innovate at the highest level and bring up their game a notch.

Open Source Platform

Android is an open source platform. Most of the stack, from low-level native, Dalvik virtual machine, application framework, and standard apps, is totally open.

Aside from the Linux kernel itself, Android is licensed under business-friendly licenses (Apache/MIT/BSD) so that others can freely extend it and use it for variety of purposes. Even some third-party open source libraries that were brought into the Android stack were rewritten under new license terms.

So, as a developer, you have access to the entire platform source code. This allows you to see how the guts of the Android operating system work. As a manufacturer, you can easily port Android OS to your specific hardware. You can also add your own proprietary secret sauce, and you do not have to push it back to the development community if you don't want to.

There's no need to license Android. You can start using it and modifying it today, and there are no strings attached. In addition, Android has many hooks at various levels of the platform, allowing anyone to extend it in unforeseen ways.

 There are couple of minor low-level pieces of code that are proprietary to each vendor, such as the software stack for the cellular, WiFi, and Bluetooth radios. Android tries hard to abstract those components with interfaces so that vendor-specific code can be managed easily.

Designed for Mobile Devices

Android is a purpose-built platform for mobile devices. When designing Android, the team looked at which mobile device constraints likely were not going to change for the foreseeable future. For one, mobile devices are battery powered, and battery performance likely is not going to get much better anytime soon. Second, the small size of mobile devices means that they will always be limited in terms of memory and speed.

These constraints were taken into consideration from the get-go and were addressed throughout the platform. The result is an overall better user experience.

Android was designed to run on all sorts of physical devices. Android doesn't make any assumptions about a device's screen size, resolution, chipset, and so on. Its core is designed to be portable.

History

The history of Android is interesting and offers some perspective on what the future might hold.

These are the key events of the past few years:

- In 2005, Google buys Android, Inc. The world thinks a "gPhone" is about to come out.
- Everything goes quiet for a while.
- In 2007, the Open Handset Alliance is announced. Android is officially open sourced.
- In 2008, the Android SDK 1.0 is released. The G1 phone, manufactured by HTC and sold by the wireless carrier T-Mobile USA, follows shortly afterward.
- 2009 sees a proliferation of Android-based devices. New versions of the operating system are released: Cupcake (1.5), Donut (1.6), and Eclair (2.0 and 2.1). More than 20 devices run Android.
- In 2010, Android is second only to BlackBerry as the best-selling smart phone platform. Froyo (Android 2.2) is released and so are more than 60 devices that run it.
- In 2011, Android is the #1 mobile platform by number of new activations and number of devices sold. The battle for dominating the tablet market is on.
- In 2012, GoogleTV, powered by Android and running on Intel x86 chips, is released. Android is now running on everything from the smallest of screens to the largest of TVs.
- In 2013, Google Glass, a wearable computing platform with an optical head-mounted display powered by Android is released to a select few.
- Beyond phones, tablets, and TVs, Android continues to be the big challenger to Embedded Linux as the platform for developing a number of specialized devices, such as home automation systems, car dashboards and navigation systems, as well as NASA satellites.

In 2005, when Google purchased Android, Inc., the world thought Google was about to enter the smartphone market, and there were widespread speculations about a device called the gPhone.

Google's CEO, Eric Schmidt, made it clear right away that Android's ambitions were much larger than a single phone. Instead, Android engineers envisioned a platform that would enable many phones and other devices.

Google's Motivation

Google's motivation for supporting the Android project seems to be having Android everywhere and by doing that, creating a level playing field for mobile devices. Ultimately, Google is a media company, and its business model is based on selling advertising. If everyone is using Android, then Google can provide additional services on top of it and compete fairly. This is unlike the business models of other software vendors who depend on licensing fees.

Although Google does license some proprietary apps, such as Gmail and Google Maps, and continues to make money off its Google Play service, its primary motivation is still the advertising revenue that those apps bring in.

As Android growth and stiff competition continue to take even Google by surprise, it is going to be essential to keep Android open for others to "remix" it in whatever way they see fit.

Android Compatibility

From the get-go, Google created Compatibility Test Suite (CTS), defining what it means to be an Android-compatible device. CTS is a combination of automated tests as well as a document that specifies what an Android device must have, should have, or what features are simply optional.

The goal of CTS is to ensure that, for a regular consumer, an average app from the market will run on an average Android device if that device claims to be supporting a certain version of Android. It is designed to prevent so-called fragmentation of the Android operating system, such as the one that happened in the world of Linux desktops, for example.

The issue with CTS is that it is up to the creator of a custom Android version to self-test its compatibility. It appears that the only major "teeth" in enforcing CTS on the part of manufacturers is Google itself, by simply not wanting to license its proprietary Android code to noncompatible devices. That proprietary code includes Google Play, Gmail, Google Maps, and much more.

CTS helps to shield the average Joe from being disappointed by an app not running as advertised due to lack of features on his device. However, CTS is by no means a must.

For example, Amazon has released Kindle Fire, a device built on top of the Android OS. Kindle Fire was never designed with CTS in mind—Amazon simply wanted a great ebook reader and saw in Android an open platform that would get it there faster.

This is a good thing, and hopefully the future of Android will stay compatible for an average Android-branded device, yet open for custom purpose-built gadgets that want to leverage its powerful software stack.

Note that manufacturers by no means have to adhere to CTS. Anyone is welcome to download and "remix" Android in any way they see fit. And people do: Android has been purpose-customized for everything from cars to satellites, and from photocopiers to washing machines. The major reason manufacturers would want to ensure Android compatibility is access to Google Play, and its rich set of apps.

Open Handset Alliance

For this to be bigger than just Google, Android is owned by the Open Handset Alliance, a nonprofit group formed by key mobile operators, manufacturers, software companies, and others. The alliance is committed to openness and innovation for the mobile user experience.

In practice, the alliance is still very young and many members are still learning to work with one another. Google happens to be putting the most muscle behind the Android project at the moment.

Android Versions

Like any software, Android is improved over time, which is reflected in its version numbers. However, the relationship between different version numbers can be confusing. Table 1-1 helps explain that.

Table 1-1. Android OS platform versions

Android version	API level	Codename
Android 1.0	1	
Android 1.1	2	
Android 1.5	3	Cupcake
Android 1.6	4	Donut
Android 2.0	5	Eclair
Android 2.01	6	Eclair
Android 2.1	7	Eclair
Android 2.2	8	Froyo (frozen yogurt)
Android 2.3	9	Gingerbread
Android 2.3.3	10	Gingerbread

Android version	API level	Codename
Android 3.0	11	Honeycomb
Android 3.1	12	Honeycomb
Android 3.2	13	Honeycomb
Android 4.0	14	Ice Cream Sandwich
Android 4.0.3	15	Ice Cream Sandwich
Android 4.1	16	Jelly Bean
Android 4.2	17	Jelly Bean
Android 4.3	18	Jelly Bean
Android 4.4	19	KitKat

The Android version number itself partly tells the story of the software platform's major and minor releases. What is most important is the API level. Version numbers change all the time, sometimes because the APIs have changed, and other times because of minor bug fixes or performance improvements.

As an application developer, you will want to make sure you know which API level your application is targeting in order to run. That API level will determine which devices can and cannot run your application.

Typically, your objective is to have your application run on as many devices as possible. So, with that in mind, try to shoot for the lowest API level possible. Keep in mind the distribution of Android versions on real devices out there. Figure 1-1 shows a snapshot of the Android Device Dashboard (*http://bit.ly/X3KDsh*) from mid-2013.

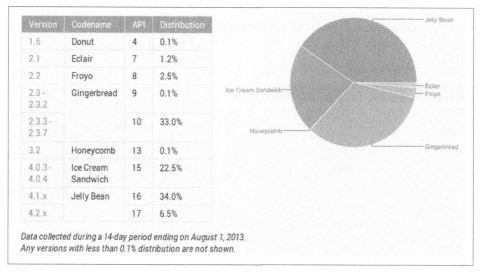

Figure 1-1. Historical Android version distribution through August 2013

You may notice that there are a lot of users of Android 2.3.3+ and 4.1.x. This places the latest and greatest (4.1.x) version as the second largest version currently in the wild. This hasn't always been the case because OEMs tended to be very slow in upgrading their OS versions. However, this has changed with Google's strong push to get everyone onto the latest and greatest. Unfortunately, there are still a lot of people who have the 2.3.3 version because they have yet to upgrade their phones to a phone with the hardware capable of handling the newer version. This is changing now because people can upgrade their phones automatically when they renew their plans.

With that in mind, you will probably choose 2.3.3 as your minimum development target, unless you truly need the features of the latest version.

Android Flavors

Android is open, and as such, many parties download it, modify it, and release their own flavors of it. Let's take a look at the options in this space.

Android Open Source Project

The official version of Android, the one that comes from Google, is technically referred to as Android Open Source Project, or AOSP for short. Think of AOSP as a reference version of Android, a vanilla flavor. You may rarely find AOSP version on a consumer device. It is often spiced up, or mixed in with some other flavors to create a better overall experience.

Manufacturer Add-Ons

Before Android, many original equipment manufacturers (OEMs) used to have teams of engineers working on low-level components of the OS that they now get for free with Android. So they started differentiating their devices by moving the innovation from reinventing the wheel to much higher-level components that their users desire. This has opened up a revolution of innovation in the mobile space.

Companies such as HTC, Motorola, and Samsung often add many useful features to vanilla Android. These additional features are sometimes referred to as overlays, skins, or mods, and are officially known as add-ons.

Some add-ons may be simple changes in the set of applications shipped with that version of Android. Others may be much more profound overhauls of the entire Android stack, such as in HTC Sense.

Often, these modification still adhere to Android Compatibility Test Suite, and make for a better user experience. Overall, the add-ons showcase the power of an open operating system and, as such, are very welcome in pushing mobile computing to the next level.

Summary

The Android operating system was designed from the ground up to be a comprehensive open source platform for mobile devices. It is a game changer in the industry and has enjoyed great success.

In Chapter 3, we'll take a look at the entire Android operating system at a high level to gain a technical understanding of how all the pieces fit together.

## CHAPTER 2

Java Review

The purpose of this chapter is to do a quick review of Java syntax and concepts. This is not in any way a true in-depth introduction to Java (for that we suggest Oracle's Java Tutorial (*http://docs.oracle.com/javase/tutorial*)). Rather, the intention is to provide a quick run-through from the very basics to more complex concepts that you will definitely need to be comfortable with in order to program for the Android platform. It is assumed that you have installed the Java Development Kit (JDK) 1.6 on the development machine (see Chapter 4 to install the JDK).

As with all opening examples for most languages, let us first cover the basic Java program and its execution with the classic "Hello World" example:

1. Open up a text editor and add the code as shown in Example 2-1.

2. Save this file as *HelloWorld.java*.

3. As shown in Example 2-2, compile using the `javac` command at a command prompt. This should create a file called *HelloWorld.class*.

4. Then using the `java` command (Example 2-2), execute the program.

5. The output should look like Example 2-3.

Example 2-1. Hello World

```
public class HelloWorld {

        public static void main(String[] args) {

                System.out.println("Hello World");

        }
}
```

Example 2-2. Hello World compile and execute

```
javac HelloWorld.java

java HelloWorld
```

Example 2-3. Hello World output

```
Hello World
```

At this stage we are looking at a really basic program that does not get into any of Java's syntax and utility. It does contain within it the three main elements of Java: a class, a variable, and a method. A variable can be considered a noun, and a method can be considered a verb. Both are parts of a class. The method that is contained is the line `public static void main(String[] args)`. This main method is what the Java run-time system calls (it's an entry point, so to speak). Without this method in the class that is specified to the Java interpreter, an error occurs and the program terminates. Now onto the simple example, *SimpleExample.java*:

1. Open up a text editor and add the code as shown in Example 2-4.
2. Save this file as *SimpleExample.java*.
3. As shown in Example 2-5, compile using the `javac` command at a command prompt.
4. Then using the `java` command (Example 2-5), execute the program.
5. The output should look like Example 2-6.

Example 2-4. A simple example

```
package com.marakana.examples;

/*
    A simple example class with a basic main method
    that instantiates several objects of the class,
    manipulates the objects, and outputs information
    about the object
*/
public class SimpleExample {

        private int number;

        public SimpleExample() { }

        public void setValue(int val) {
                number = val;
        }

        public int getNumber() {
                return number;
```

```
        }

        public static void main(String[] args) {
                for(int i=0;i<10;i++) {
                        SimpleExample example = new SimpleExample();

                        if(i/2 <= 2) {
                                example.setValue(i);
                        } else {
                                example.setValue(i*10);
                        }

                        System.out.println("SimpleExample #"+i+
                                "'s value is "+example.getNumber());
                }
        }
}
```

Example 2-5. SimpleExample compile and execute

```
javac -d . SimpleExample.java

java -cp . com.marakana.examples.SimpleExample
```

Example 2-6. SimpleExample sample output

```
SimpleExample #0's value is 0
SimpleExample #1's value is 1
SimpleExample #2's value is 2
SimpleExample #3's value is 3
SimpleExample #4's value is 4
SimpleExample #5's value is 5
SimpleExample #6's value is 60
SimpleExample #7's value is 70
SimpleExample #8's value is 80
SimpleExample #9's value is 90
```

Note the use of the -d parameter with the `javac` command, which tells the compiler that the destination of the compiled class (`SimpleExample`) and its directory structure's root is the local directory in which the Java file is located. What this means is that a directory named *com* will be created. Within this *com* directory, an *examples* directory is placed, and within *examples*, *SimpleExample.class* is generated (see Figure 2-1). This structure follows that of the `package com.marakana.examples;` line dictated at the top of the Java file. The purpose of this packaging structure is to avoid collision of class names. For example, "com.marakana.examples.SimpleExample" and "org.samples.SimpleExample" are both classes named "SimpleExample" but they reside in different namespaces. This naming structure tends to follow these conventions:

- Package names are all lowercase.

- Packages in the Java language begin with "java" or "javax."

- Generally, companies use their Internet domain in reverse order (so a company like *oreilly.com* would become *com.oreilly*, *nonprofit.org* would become *org.nonprofit*, etc.). If the domain contains some special characters (nonalphanumeric) or conflicts with a reserved Java keyword, it is either not used or an underscore (_) is used instead.

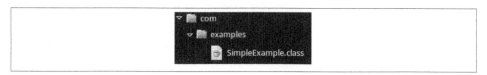

Figure 2-1. Package tree

It is this package naming scheme that is used when executing the program (i.e., "com.marakana.examples.SimpleExample"). With the `java` command, the `-cp` (class-path) option is used to designate where the command should seek out the specified class(es). In Example 2-5 . is used to designate that the root directory for the classes is the current local directory. The `java` and `javac` commands have a variety of other options that are useful to check out.

Now that we have an example that runs and is a bit more substantial, let's dive into some of the specifics.

Comments

Comments are sections in the code that are either explanatory or contain code that is not intended to execute. Comments are expressed in one of two ways: either with `//` to denote a single-line comment, or with `/* */` to denote a multiline comment (see Example 2-7). The single-line case dictates that everything on that line to the right of `//` is a comment. The multiline case spans from the `/*` (everything to the right of it) to the left of `*/`.

Example 2-7. Comments

```
// This is an example of a single line comment

/* This is an example
   of a multi line
   comment */
```

Data Types: Primitives and Objects

Java is an object-oriented, statically typed language. *Object-oriented* is a programming paradigm that is based on the concept of *objects*. This idea is often analogous to that of the real world, where we have things (such as cars and people) and the things have properties (such as doors and legs) and behavior/actions (such as turning right and walking). What *statically typed* means is that Java checks the declaration of the data type of every variable in the program at compile time. This enforcement of the data type ensures that variables cannot change what they mean within the program once they have been declared (e.g., a number cannot be swapped for text or vice versa). The types of data fall into two camps: primitive data types and objects.

Following are the eight primitive data types in Java:

boolean
> 1-bit true (1) or false (0) value

byte
> 8-bit signed whole number (no decimals) with values ranging from −128 to 127

short
> 16-bit signed whole number with values ranging from −32,768 to 32,767

int
> 32-bit signed whole number with values ranging from −2,147,483,648 to 2,147,483,647

long
> 64-bit signed whole number with values ranging from −9,223,372,036,854,775,808 to 9,223,372,036,854,775,807

float
> Single-precision 32-bit floating-point number (has decimal)

double
> Double-precision, 64-bit floating-point number

char
> A single, 16-bit Unicode character; for example, the letter "A" (note that "a" and "A" are different characters)

The other data type that everything else falls under is an object. An object is a complex type in that within each object are a variety of properties (also called *fields* or *variables*) and methods (also called *functions*). All objects are defined by a blueprint called a *class* (making objects an *instance* of a class). In many cases, the class is described as a file with the file extension of *.java* (such as *SimpleExample.java*) and is compiled into a machine-readable file with the file extension *.class* (see Example 2-5).

Taking a look at the *SimpleExample.java* class (Example 2-4), there is only one variable that is declared within it: number (private int number). This integer number is not expressly assigned and so by default is set to 0 (all number cases default to the value 0 respective to their type). In the case of a boolean, the default value is false, whereas in the case of char it is \u0000 (in other words, zero expressed as a UNICODE value). If a variable is an object and it was not assigned anything, the default value would be something called null. null is a special value that means "not assigned" or "unknown" (it's a bit more complex than that, but we are trying to keep things simple).

Continuing on with the example, there is a method called setValue() that takes in int (integer) as its input and then sets the number variable to that integer. To access the value of the number variable, another method is declared called getNumber() that returns the number variable. These two are examples of what a typical method declaration may look like. A method declaration is made up of six pieces:

Modifier
> This defines the access type (e.g., public) and kind (e.g., static).

Return type
> This defines the data type that is returned (e.g., int). If no data type is to be returned, void is used.

Name
> The name of the method.

Input parameter list
> A comma-delimited list of parameters preceded with their data type (e.g., int val, String str, double num).

Exception list
> A comma-delimited list of exception types that are thrown by the method.

Body
> The logic/code between the braces of the method.

There is what looks like a method using the name of the class but does not have a return type (public SimpleExample()). This is called a *constructor*—its purpose is to enable object instances of the class to be instantiated via the use of the *new* operator (SimpleExample example = new SimpleExample()).

Modifiers

Modifiers are split into two categories: access modifiers (public, protected, private, and no-access modifier) and nonaccess modifiers (static, final, strictfp, abstract, synchronized, volatile, transient, and native). We will cover the access modifiers and only one of the nonaccess modifiers, static.

Access modifiers define the level of access a method or variable has. It is a type of security in that a hierarchy of control is established when using them:

public
> Everyone can see and access this code.

protected
> The class this is defined in, classes within the same package, or classes that are subclasses of the class this is defined in can view and access this code.

default/nonaccess modifier
> The class this is defined in and classes within the same package can view and access this code.

private
> Only the class that this is contained within can see and access this code.

When an object is instantiated from the class blueprint, it has a distinct copy of the instance variables of its own. With the use of the *static* modifier, the variable is associated directly with the class and only one is ever created. This static variable becomes a common variable across the object's instances of the class. In the same way, static methods are only accessible at the class level. For example, the main method (`public static void main(String[] args)`), ensures that only one main method exists for this class and exists as an entry point to the execution of the program.

Arrays

Looking at the main method's input, we have the variable `args`. `args` is a string array (`String[] args`) denoted with the `[]`. An array is a container object that holds a fixed number of values of a specific type—in essence, it is a list of values. The declaration of the array sets the type that is held within each element of the array, and the size is fixed when it is assigned (see Example 2-8). Each element of the array is accessed by its numerical index, which is a number representing where it is located in the ordered list. Note that the index starts at 0 and increments up by 1 until the last element is one less than the total size.

Example 2-8. Array declarations and value assigning

```
double[] someArray;   // declaring

someArray = new double[4];  // assigning size of 4

int[] integerArray = new int[10];  // declaring and assigning size of 10
```

```
integerArray[0] = 32;   // assigning the first element
integerArray[1] = 12;
integerArray[2] = 333;
integerArray[3] = 3343;
integerArray[4] = 1;
integerArray[5] = 99;
integerArray[6] = 42;
integerArray[7] = -33;
integerArray[8] = 32;
integerArray[9] = 0;   // assigning the last element

// another way to declare and assign

    // declaring and assigning 3 elements directly
String[] anotherArray = {"Some String","a","strings"}
```

Operators

Operators are special characters that denote actions performed on a variable. Such operators include things such as basic math, boolean logic, and assignment. The operators have an order of hierarchy as to what operations are done first. The following shows a table of operators in order of priority:

```
| postfix               | expr++ expr--                    |
| unary                 | ++expr --expr +expr -expr ~ !    |
| multiplicative        | * / %                            |
| additive              | + -                              |
| shift                 | << >> >>>                        |
| relational            | < > <= >= instanceof             |
| equality              | == !=                            |
| bitwise AND           | &                                |
| bitwise exclusive OR  | ^                                |
| bitwise inclusive OR  | |                                |
| logical AND           | &&                               |
| logical OR            | ||                               |
| ternary               | ? :                              |
| assignment            | = += -= *= /= %= &= ^= |= <<= >>= >>>=|
```

Control Flow Statements

Moving on to more complex forms of code logic, we now discuss control flow statements. Control flow statements are blocks of code that break up the flow of execution (the main flow being top to bottom) and provide a means for branching and looping.

The simplest control flow statements are if-then and if-then-else.

The SimpleExample program in Example 2-4 contains the following section of code:

```
if(i/2 <= 2) {
        example.setValue(i);
```

```
    } else {
        example.setValue(i*10);
    }
```

This describes the logic of "IF current value of i divided by 2 is less than or equal to 2, THEN call the method setValue on the example object and pass in the current value of i, ELSE call the method setValue on the example object and pass in the current value of i times 10." As you can see, the point of the if-else type control statement is to create decision points based on states within the code.

Another control statement that is very similar to if-else is the switch statement (see Example 2-9). The switch statement provides multiple execution paths that depend on the conditions of the state.

Example 2-9. Switch case

```
int somenumber = 0;

// some logic changes somenumber's value making it either 0, 1, or 2

switch(somenumber) {
    case 0: doSomething();
        break;
    case 1: doSomethingOne();
        break;
    case 2: doSomethingTwo();
        break;
}
```

The next type of control statement is the loop. There are four types of loops within Java (see Example 2-10). The first is the while loop, which executes the code within its block so long as the input (expression) to the while statement's state is true. The second loop is the do-while loop. This differs from the while loop in that the block of code within the do portion is executed first, and then the expression within the while portion is checked. This ensures that the code in the do portion executes at least once. Next is the for loop. The for loop executes the loop until a condition defined within the loop's input is met. This enables the programmer to create a conditional and incremental loop. Lastly, the for-each loop provides a quick and easy way for the programmer to iterate through a variable list.

Example 2-10. Different loops

```
//-- the while loop

int i = 0;

// until i is equal or greater than 10 does the loop continue
while(i < 10) {
    System.out.println(String.valueOf(i));
    i++;
```

```
}

//-- the do while loop

int k = 0;

// until k is equal or greater than 10 does the loop continue
do {
        System.out.println(String.valueOf(k));
        k++;
} while(k < 10);


//-- the for loop

// the loop initializes j to 0, then increments it by +1
// (the j++) until j is equal to or greater than 10
for(int j=0;j<10;j++) {
        System.out.println(String.valueOf(j));
}

//-- for each loop

String[] arr = {"The","Quick","Brown","Fox"};

// the loop iterates through the entire array
for(String a: arr) {
        System.out.println(a);
}
```

The final control flow statements are the branching statements: break, continue, and return (see Example 2-11). A break statement terminates the most innermost loop or switch statement it is in. The continue statement causes a skip ahead (to the next iteration, thus skipping only the current one) to occur within a loop. The return statement exits from the current method and may or may not pass a value.

Example 2-11. Break, continue, and return in a loop

```
        // forloop1
for(int i=0;i<10;i++) {

        // if i is even then continue to the next iteration of forloop1
  if(i%2 == 0) continue;
  else {
        // forloop2
        for(int j=0;j<5;j++) {
        // if j%i has no remainder then jump out of
        //   forloop2 and back to forloop1

            if(j%i != 0) break;
            else return i;
```

```
        // else return the integer value i
        //    and then stops the complete flow
        }
    }
}
```

Error/Exception Handling

We have now covered the basics shown in the *SimpleExample.java* case. Before we launch into some more complex subjects, we need to cover the concept of error/exception handling. To do this, we can take the SimpleExample case and add to it to do some error handling as shown in Example 2-12. The resulting output would look like Example 2-13.

Example 2-12. SimpleExample with error handling

```
package com.marakana.examples;

/*
    A simple example class with a basic main method
    that instantiates several objects of the class,
    manipulates the objects, and outputs information
    about the object
*/
public class SimpleExampleWErrorHandling {

    private int number;

    public SimpleExampleWErrorHandling() { }

    //------- ERROR HANDLING PART 1
    public void setValueWithException(int val) throws Exception {
        if(val < 0) throw new Exception(
            "setValue Exception- Value that is set is Negative!");
        number = val;
    }

    public int getNumber() {
        return number;
    }


    // here we override toString so the set value
    // is returned rather than the object's textual
    // representation

    @Override
    public String toString() {
        return value;
    }
```

```java
    public static void main(String[] args) {
        for(int i=0;i<10;i++) {
            SimpleExample example = new SimpleExample();

            if(i/2 <= 2) {
                    //------- ERROR HANDLING PART 4
                try { example.setValue(i); }
                catch (Exception e) { e.printStackTrace(); }
            } else {
                    //------- ERROR HANDLING PART 4
                try { example.setValue(i*10); }
                catch (Exception e) { e.printStackTrace(); }
            }

            System.out.println("SimpleExample #"+i+
                    "'s value is "+example.getNumber());
        }


        showErrorHandling(); //------- ERROR HANDLING PART 2

    }

    //------- ERROR HANDLING PART 3
    public static void showErrorHandling() {
        // here we show Error Handling
      try {
        System.out.println();
        System.out.println("SimpleExample BadValue Insert Case Start");
        SimpleExample example = new SimpleExample();
        example.setValueWithException(-10);
        System.out.println("SimpleExample BadValue's value is "+
                              example.getNumber());
        System.out.println("SimpleExample BadValue Insert Case End");

      } catch (Exception e) {

        System.err.println("SimpleExample BadValue "+
                              "Insert Case threw an exception");
        e.printStackTrace();

      } finally {

        System.out.println("SimpleExample BadValue "+
                              "Insert Case Finally Called");

      }
    }
}
```

Example 2-13. SimpleExample with error handling output

```
SimpleExample #0's value is 0
SimpleExample #1's value is 1
SimpleExample #2's value is 2
SimpleExample #3's value is 3
SimpleExample #4's value is 4
SimpleExample #5's value is 5
SimpleExample #6's value is 60
SimpleExample #7's value is 70
SimpleExample #8's value is 80
SimpleExample #9's value is 90

SimpleExample BadValue Insert Case Start
SimpleExample BadValue Insert Case threw an exception
java.lang.Exception: setValue Exception- Value that is set is Negative!
        at com.marakana.examples.SimpleExampleWErrorHandling.setValue
    (SimpleExample.java:17)
        at com.marakana.examples.SimpleExampleWErrorHandling.showErrorHandling
    (SimpleExample.java:51)
        at com.marakana.examples.SimpleExampleWErrorHandling.main
    (SimpleExample.java:40)
SimpleExample BadValue Insert Case Finally Called
```

An exception is an event that disrupts the normal flow of program execution. This can be deemed as an *error* because it breaks from the normal flow. When the error occurs, an object, called an exception object, is generated with information about the error and is passed to the runtime system. Creating an exception and passing it to the runtime system is called "throwing an exception."

In Example 2-12, a comment stating ERROR HANDLING PART 1 is right above the method called setValue(). Here the original method changed to declare the terms "throws Exception." This states that the method could throw an exception object of class Exception (a variety of subclasses could be specified, such as IOException). The logic in this method has also been changed. Should the input value be negative, the logic explicitly instantiates an Exception object and then "throws" it (throw new Exception(""set Value Exception- Value that is set is Negative!"")).

The comment stating ERROR HANDLING PART 2 refers to the method that is referenced below the ERROR HANDLING PART 3 comment. This method, showErrorHandling(), contains within its body the mechanism to handle the error/exception. The mechanism is the try-catch-finally block. Code within the "try" section is covered in that should an exception get thrown, and the exception type is the same class or a subclass of the exception type that is defined in the catch, then the catch's body is executed. Note that the lines of code after the method call that throws the exception never get executed. Whether or not an exception is thrown, the finally block's code will always get executed after the catch or try completes. Note that because setValue() now throws an exception, we had to wrap the other setValue calls in the main() method with try-catch blocks (as shown by the ERROR HANDLING PART 4 sections).

Complex Example

This section walks through a series of examples to illustrate some of the more complex topics related to Android programming:

1. Open up a text editor and copy and paste six files: Example 2-14, Example 2-15, Example 2-16, Example 2-17, Example 2-18, and Example 2-19.

2. As shown in Example 2-20, compile using the javac command at a command prompt.

3. Using the java command (Example 2-20), execute the program.

4. The output should look something like Example 2-21 (the output will vary because there is a random element in play).

Example 2-14. Complex Example—ComplexExample.java

```java
package com.marakana.examples;

/*
  In this example, ComplexExample has a main method when executed
  instantiates a MsgGenerator object and then passes this object to a
  Thread.  The Thread's process is then started and then the main thread
  waits till the generator object notifies that it is done (via the
  notifyAll()).  At this point the generator's printList method is called
  and information about what was in the generator's list is printed out.
*/
public class ComplexExample {

        public static void main(String[] args) {
                System.out.println("ComplexExample start");

                MsgGenerator generator = new MsgGenerator();

                Thread thread = new Thread(generator);
                thread.start();

                try {
                        synchronized(generator) {
                                generator.wait();
                        }
                } catch (InterruptedException ie) {
                        System.err.println("Generator Wait Interrupted!!!");
                        ie.printStackTrace();
                } finally {
                        System.out.println("Generator Wait End");
                }
```

```
                    generator.printList();

                    System.out.println("ComplexExample end");
            }
}
```

Example 2-15. ComplexExample—MsgInterface.java

```
package com.marakana.examples;

public interface MsgInterface {
        void setMsg(String msg);
        String getMsg();
        String getMsgType();
}
```

Example 2-16. ComplexExample—MsgTypeImplementation.java

```
package com.marakana.examples;

public class MsgTypeImplementation implements MsgInterface {

        private String msg;

        public void setMsg(String msg) {
                this.msg = msg;
        }

        public String getMsg() {
                return this.msg;
        }

        public String getMsgType() {
                return "MsgTypeImplementation";
        }
}
```

Example 2-17. ComplexExample—MsgTypeAdditional.java

```
package com.marakana.examples;

public class MsgTypeAdditional implements MsgInterface {

        private String msg;

        public MsgTypeAdditional() { }

        public MsgTypeAdditional(String msg) {
                setMsg(msg);
        }

        public void setMsg(String msg) {
```

```
            this.msg = msg + " 2";
    }

    public String getMsg() {
            return this.msg;
    }

    public String getMsgType() {
            return "MsgTypeAdditional";
    }
}
```

Example 2-18. ComplexExample—MsgTypeOneExtended.java

```
package com.marakana.examples;

public class MsgTypeImplementationExtended extends MsgTypeImplementation {

    @Override
    public String getMsgType() {
            return "MsgTypeImplementationExtended";
    }


    // here we Overload the getMsg() method so we now have an
    // additional method that adds something to the getMsg() string

    public String getMsg(String pre) {
            return pre+" "+getMsg();
    }

    // and again we Overload the getMsg() method this time with a int

    public String getMsg(int post) {
            return " -- "+post;
    }
}
```

Example 2-19. ComplexExample—MsgGenerator.java

```
package com.marakana.examples;

import java.util.ArrayList;
import java.util.Random;

public class MsgGenerator implements Runnable {
    private ArrayList<MsgInterface> list;

    public MsgGenerator() {
            list = new ArrayList<MsgInterface>();
    }

    public void run() {
```

```
            Random rand = new Random();
            int r = 0;
            ArrayList<MsgInterface> localList =
                            new ArrayList<MsgInterface>();

            while((r = rand.nextInt(20)) < 18) {
                    MsgInterface msg = null;

                    switch (rand.nextInt(3)) {
                            case 0: msg = new MsgTypeImplementation();
                                    break;
                            case 1: msg = new MsgTypeAdditional();
                                    break;
                            case 2: msg =
                                    new MsgTypeImplementationExtended();
                                    break;
                    }

                    msg.setMsg("Num is: "+r);

                    localList.add(msg);
            }

            synchronized(this) {
                    list = localList;
                    this.notifyAll();
            }
    }

    public void printList() {
        ArrayList<MsgInterface> localList;
            synchronized (this) {
                localList = list;
            }

        System.out.println("List Contents:");
        for(MsgInterface msg : localList) {
        System.out.println("  "+msg.getMsgType()+" msg = "+msg.getMsg());
          if(msg.getMsgType().equals("MsgTypeImplementationExtended")) {
            System.out.println(" *** Overloaded getMsg : "+
            ((MsgTypeImplementationExtended) msg).getMsg("Special") +
            ((MsgTypeImplementationExtended) msg).getMsg(99));
          }
        }
        System.out.println("List Size: "+list.size());
    }
}
```

Compile using the `javac` command in a command prompt, and then use the `java` command to execute the program, as shown in Example 2-20.

Example 2-20. SimpleExample compile and execute

```
javac -d . ComplexExample.java MsgInterface.java MsgTypeImplementation.java \
        MsgTypeAdditional.java MsgTypeImplementationExtended.java \
        MsgGenerator.java
```

OR

```
javac -d . *.java
```

```
java -cp . com.marakana.examples.ComplexExample
```

The output should look something like Example 2-21 (the output will vary because there is a random element in play).

Example 2-21. ComplexExample sample output

```
ComplexExample start
Generator Wait End
List Contents:
  MsgTypeAdditional msg = Num is: 2 2
  MsgTypeAdditional msg = Num is: 7 2
  MsgTypeImplementationExtended msg = Num is: 13
 *** Overloaded getMsg : Special Num is: 13 -- 99
  MsgTypeImplementation msg = Num is: 13
  MsgTypeImplementationExtended msg = Num is: 6
 *** Overloaded getMsg : Special Num is: 6 -- 99
  MsgTypeImplementationExtended msg = Num is: 2
 *** Overloaded getMsg : Special Num is: 2 -- 99
  MsgTypeAdditional msg = Num is: 2 2
  MsgTypeAdditional msg = Num is: 16 2
  MsgTypeAdditional msg = Num is: 9 2
  MsgTypeImplementation msg = Num is: 15
  MsgTypeImplementation msg = Num is: 13
  MsgTypeImplementationExtended msg = Num is: 2
 *** Overloaded getMsg : Special Num is: 2 -- 99
  MsgTypeImplementationExtended msg = Num is: 6
 *** Overloaded getMsg : Special Num is: 6 -- 99
  MsgTypeImplementation msg = Num is: 10
  MsgTypeAdditional msg = Num is: 10 2
  MsgTypeImplementationExtended msg = Num is: 10
 *** Overloaded getMsg : Special Num is: 10 -- 99
List Size: 16
ComplexExample end
```

Interfaces and Inheritance

The ComplexExample program in the preceding section used multiple classes. Of these, look closely at MsgTypeImplementation, MsgTypeAdditional, and MsgTypeImplementationExtended. Both MsgTypeImplementation and MsgTypeAdditional use the term *implements* and reference MsgInterface. The file *MsgInterface.java*, unlike the other files, uses the term *interface* rather than *class*. An interface is a reference type similar to a class, but only contains within it constants, method signatures, and nested types. In the MsgInterface case we only have method signatures, which are skeleton descriptions of methods (name, return type, and argument types are described). MsgTypeImplementation and MsgTypeAdditional implement MsgInterface; they are fleshed-out versions of MsgInterface. By doing this, both classes must have defined within them the methods described in MsgInterface. What this does is enable the code in MsgGenerator to view instances of MsgTypeImplementation and MsgTypeAdditional as instances of MsgInterface. Note that classes may implement multiple interfaces, thus being perceived as multitypes.

The class MsgTypeImplementationExtended uses the term *extends* and then references MsgTypeImplementation. This is inheritance, where one class is a subclass of another. Unlike the case where a class may implement multiple interfaces, only one class may be extended. Thus, there is a clear chain of parent-to-child-class that is defined. In this case, because MsgTypeImplementationExtended is a subclass of MsgTypeImplementation, it too is a class that implements MsgInterface. However, because MsgTypeImplementationExtended is a subclass of MsgTypeImplementation, and MsgTypeImplementation has defined the methods that MsgInterface described, MsgTypeImplementationExtended has no need to define the methods. It can, however, override a method and make it its own, as in the case of getMsgType.

One other thing that we have done within MsgTypeImplementationExtended is to create two methods similar to the basic getMsg() that all MsgInterfaces must have: getMsg(String pre) and getMsg(int post). Note that the return type is the same (String) and the method name is the same (getMsg), but we have included an input variable (String pre and int post). This is called *overloading* and is specifically defined as the ability to have more than one method with the same name in a class. What distinguishes them is the difference in the parameter list (inputs). In MsgTypeImplementationExtended, both getMsg() and getMsg(String pre) can coexist and the compiler is able to figure out which method to call based on the inputs passed. Notice that in the printList() method of MsgGenerator, we see if the MsgInterface in question is a MsgTypeImplementationExtended class ((MsgTypeImplementationExtended) msg) and then we call both the getMsg(String pre) and getMsg(int post) methods. We have to *cast* the msg object as a MsgTypeImplementationExtended in order to call the two methods because msg is initially referenced as a MsgInterface, which does not have the two methods in question.

Collections

In this example, MsgGenerator has a variable called a *list*. This list is an instance of a special class called ArrayList. ArrayList is part of the *java.util* package and is part of a group of classes called *collections*. A collection is an object that groups multiple objects into a single unit. In this case, `ArrayList` is a list of objects that are kept in a specific order and each object may be referenced by an index (much like an array). It is a very good idea to become familiar with the collections type such as `Map` (key-value paired set of objects) and `Set` (a group of objects that are guaranteed to be unique within the set).

Generics

The `ArrayList` defined in `MsgGenerator` references `<MsgInterface>`. This is the use of *generics*. Though a full description of generics is out of scope for this review, we wish to point out the strong typing that this brings about because this is heavily relied upon throughout Android. In this case, the `ArrayList` can only hold classes that are of type `MsgInterface` (which `MsgTypeOne`, `MsgTypeTwo`, and `MsgTypeThree` all are). Any other type of object placed into the `ArrayList` will cause an exception to be thrown. This also enables the programmer to not have to cast an object that is grabbed from the `ArrayList`.

Threads

In this section we cover a very basic example of a two-threaded program. A thread is an execution process. The first thread is the main thread that is started by calling `main()`. The second thread is instantiated and started within `main()`. Here a `Thread` object is instantiated and the generator class is passed as an argument. The `MsgGenerator` class implements a interface called `Runnable`. A `Runnable` class must implement a method called `run()`. The newly created `Thread`, upon having the `start()` method called, leaps up and proceeds to run in parallel to the main thread and executes the `MsgGenerator` object's run method.

`ComplexExample`'s first thread (main) is dependent on the second thread. More specifically, the first thread ends up calling `generateList` on the generator object, which prints out to the console the list of objects and their respective message values. Because we want to get a printout of all objects in the list, we must wait for the list to get filled. However, because the two processes are running in parallel, it is difficult to figure out when the list is filled up. To find this out, there needs to be a signalling method between the two threads. In this case we use the wait-notify (`notifyAll`) method. When the main thread calls `wait` on the generator object the main thread, is effectively paused, or waiting. In the meantime, the generator object proceeds on until after the list is filled. It then calls `notifyAll`, which proceeds to let all objects waiting on it know that the list is done. At this point, the main thread stops pausing and proceeds.

In Java 5 and above, there is a whole package named *java.util.concurrent* that contains a whole set of classes dedicated toward handling and simplifying threading, such as

`ThreadPools` and `Schedulers`. It is highly recommend that you become familiar with this package.

Summary

This chapter covered a very quick review of Java, from its basic syntax to more complex topics.

## CHAPTER 3

The Stack

This chapter offers a 9,000-foot overview of the Android platform. Although you're concerned primarily with writing Android applications, understanding the layout of the system will help you understand what you can or cannot easily do with Android.

By the end of this chapter, you'll understand how the whole system works, at least from a high level. You should be able to identify each of the main layers of the platform and have a general understanding of its purpose.

Stack Overview

The Android operating system is like a cake consisting of various layers. Each layer has its own characteristics and purpose—but the layers are not always cleanly separated and often seep into one another.

As you read through this chapter, keep in mind that we are concerned only with the big picture of the entire system and will get into the nitty-gritty details later on. Figure 3-1 shows the parts of the Android stack.

Linux

Android is built on top of the Linux kernel. Linux is a great operating system, and is the poster child of open source. Its kernel has been hardened and tightened over the years by many engineers continually improving it. Many users depend on Linux every day (often unknowingly).

There are many good reasons for choosing Linux as the base of the Android stack. Some of the main ones are its portability, security, and features:

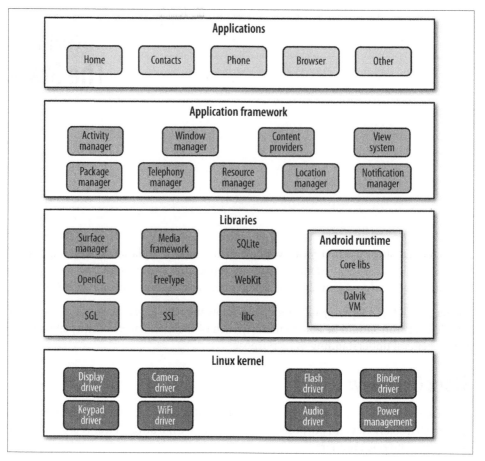

Figure 3-1. Android stack

Portability

Linux is a flexible platform that is relatively easy to port to various hardware architectures. What Linux brings to Android is a level of hardware abstraction. Because Android is based on the Linux kernel, we don't have to worry too much about underlying hardware features. Most low-level parts of Linux have been written in fairly portable C code, which allows for third parties to port Android to a variety of devices.

Security

Linux is a highly secure system, having been tried and tested through some very harsh environments over the decades. Android relies heavily on Linux for security, and all Android applications run as separate Linux processes with permissions set by the Linux system. As such, Android passes many security concerns to the underlying Linux system. Unlike some other Java-based mobile platforms, in Android the kernel is the sole enforcer of Android permissions. This allows for a simple, yet very powerful, security mechanism. It also allows Android apps access to native code, such as fast C implementations of various libraries via the Java Native Interface.

Features

The Linux kernel comes with a lot of very useful features. Android leverages many of them, such as support for memory and power management, as well as networking and radio functionality.

Android != Linux

Keep in mind that Android is not just another flavor of Linux, in the way that Ubuntu, Fedora, or Red Hat are. Many things you'd expect from a typical Linux distribution aren't available in Android, such as the X11 window manager, the ability to add a person as a Linux user (e.g., user Bob), or even the glibc standard C library.

On the other hand, Android adds quite a bit to the Linux kernel, such as an improved power management that is well-suited for mobile battery-powered devices, a very fast interprocess communication mechanism based on Binder, and a mechanism for sandboxing applications so they are isolated from one another.

 The Linux kernel is licensed under General Public License (GPL), so any modifications and additions to it must also be licensed under the same GPL open source license. Remember that Google's vision for Android is to create a platform that runs on many different devices. As such, Google expects other companies to dedicate their engineers to work on additional Android features. For that to be commercially viable, it is helpful to allow those companies to own their derivative work and be able to license it under whatever license they see fit: open or closed source. Because the GPL doesn't allow for that, Android tries hard to keep GPL code out of the rest of the Android stack. Sometimes those legal and business issues result in some interesting software architecture choices.

Native Layer

The native layer is a set of code that is written mostly in C/C++. Unlike the Linux layer, the native layer is in the so-called *user space*. This part of the stack consists of couple of different parts, such as HAL, native libraries, native daemons, and native tools.

HAL

HAL stands for hardware abstraction layer. If you recall from "Linux" on page 31, Linux was picked because of its ability to run on many various hardware boards. Indeed, Linux probably has the widest device driver support of any other operating system on the planet. The problem, however, is that access to the device drivers is usually not very standardized. That means that an application would need to know how to access a particular piece of hardware depending on the hardware manufacturer specifications and its device driver.

Android was designed to run on many different hardware configurations, and an Android app shouldn't care about specifics of certain boards. To solve this problem, Android abstracts each major device driver with a shared native library. This library is a shared object that adheres to a common interface supporting any major hardware driver. What that means is that each manufacturer needs to implement a common library and abstract out the intricacies of its specific device design.

HAL basically provides the unified device driver model that is missing in standard Linux. This is its primary role. Secondarily, it has an additional feature of keeping the GPL code out of the user space. Basically, most device drivers are implemented as Linux kernel modules, and are often built into the kernel itself. That makes them subject to GPL license rules that would possibly require any code that uses such drivers to also be licensed under GPL. As more and more people develop and customize the Android platform, having this restriction on the licensing of derived code might discourage commercial programmers, because they would be giving their intellectual property away under the same GPL rules. HAL provides a nice buffer between kernel space and the rest of the Android stack, allowing for much more flexible licensing of any derived work in the upper layers of the stack.

Native Libraries

The native libraries are C/C++ libraries. Their primary job is to support the Android Application Framework layer, which we'll explore next.

Some of these libraries are purpose-built for the Android OS, whereas others are often taken from the open source community in order to complete the operating system.

Some of the notable purpose-built Android native libraries include:

Bionic
> An Android-specific implementation of libc library, derived from the BSD project and updated for needs of Android OS. Bionic also helps keep LGPL code out of user space.

GNU libc, the default C library for Linux, is licensed under a Lesser General Public License (LGPL), which requires any changes that you release publicly to be pushed back to the open source community. As such, it might not be the most business-friendly open source license when a company wants to keep its derivative work proprietary. Bionic, on the other hand, is licensed under an Apache/MIT license, which doesn't require derivative works to be open sourced.

Binder
> A very fast inter-process communication mechanism that allows for one Android app to talk to another.

Framework libraries
> Various libraries designed to support system services, such as location, media, package installer, telephony, WiFi, voip, and so on.

Other open source libraries include:

Webkit
> A fast web-rendering engine used by Safari, Chrome, and other browsers.

SQLite
> A full-featured SQL database that the Android app framework exposes to applications.

Apache Harmony
> An open source implementation of Java libraries.

OpenGL
> 3D graphics libraries.

OpenSSL
> The secure socket layer, allowing for secure point-to-point connectivity.

Native Daemons

Native daemons are executable code that usually runs to support some kind of system service. Examples of native daemons include:

Service Manager (`servicemanager`)
> The umbrella process running all other framework services. It is the most critical native daemon.

Radio interface layer daemon (`rild`)
> Responsible for supporting the telephony functionality via GSP or CDMA, usually.

Installation daemon (`installd`)
> Supports management of apps, including installation, upgrades, as well as granting of permissions.

Media server (`mediaserver`)
> Supports camera, audio, and other media services.

Android Debug Bridge (`adbd`)
> Supports developer connectivity from your PC to the device (including the emulator) so that you can develop apps for Android.

There are about a dozen other native services. Most of these services are started by the `init` process, which we'll explore next.

Native Tools

Native tools include many standard Linux command-line tools, as well as the `init` process that is responsible for starting all the native daemons, among other things.

Like most other operating systems, Android has a command-line shell where developers can poke around the system. On Android, developers access this shell via ADB, which we'll go over later. However, if you are an experienced Linux user, you'll quickly notice that the set of commands available in the standard Android release is far smaller than other typical Linux distributions. That's because Android uses `toolbox` to support most of these command-line tools, such as `cd`, `ls`, `ps`, `top`, `df`, and so on. If you are used to Linux, do not expect to find `grep`, `vi`, `less`, `more`, or any other of the common developer tools. That's why platform developers often tend to replace the standard Android `tool box` with the Linux busybox. However, doing that is well beyond the scope of this book, because it gets into details of the Android internals.

Dalvik

Dalvik is a purpose-built virtual machine designed specifically for Android.

The Java virtual machine (VM) was designed to be a one-size-fits-all solution, and the Dalvik team felt it could do a better job by focusing strictly on mobile devices. It looked at which constraints specific to a mobile environment are least likely to change in the future. One of these is the limited battery life, and the other is the size of ever-shrinking mobile devices. Dalvik was built from the ground up to address those constraints.

To address the battery constraint, Dalvik was designed as a registry-based virtual machine, which makes it suitable for ARM-based chips. ARM tends to run much cooler than the equivalent Intel x86 type of architecture, and thus consumes less battery, which x86 chips tend to waste on heat. The standard Java VM, by comparison, is stack-based, making it suitable for today's powerful PCs and servers, most of which are plugged into the wall.

To address the size issue, Dalvik does some interesting things. When instantiating an object, the standard Java VM would locate the class file for that object on the disk and then load it into RAM. That makes sense because the disk on a typical PC or server is mechanical, thus it reads and writes at relatively slow speeds compared to RAM. Mobile devices, on the other hand, do not use hard drives but rely on solid state memory for both RAM as well as "disk" storage. To minimize doubling of limited available memory, Dalvik "loads" the class file directly on the disk, by pointing to its location. It copies into RAM only things that change, using a copy-on-write algorithm. This allows for much lower total memory usage.

Also, each Android application runs in its own process in order to provide for application sandboxing, which is the cornerstone of the Android security model. That means that at any point in time, your Android device may have a dozen or more Dalvik VMs loaded in memory. To minimize total memory consumption, Dalvik itself is made to have a tiny memory footprint, as well as to share system libraries instead of creating a copy for each instance.

Another side effect of replacing the Java VM with the Dalvik VM is the licensing. Whereas the Java language, Java tools, and Java libraries are free, the Java virtual machine is not. This was more of an issue back in 2005 when the work on Dalvik started. Nowadays, there are open source alternatives to Sun's Java VM, namely the OpenJDK (*http://openjdk.java.net/*) and Apache Harmony (*http://harmony.apache.org/*) projects. Though Android uses Apache Harmony for its Java libraries, it relies on Dalvik for the execution of the code.

By developing a truly open source and license-friendly virtual machine, Android yet again provides a full-featured platform that others are encouraged to adopt for a variety of devices without having to worry about the license.

 Dalvik was developed by Dan Bornstein and his team at Google. He named it after Dalvik, a fisherman village in Iceland. As a tribute to this virtual machine, the author got a California license plate that says DALVIK. Honk if you see it on the road!

Android and Java

In Java, you write your Java source file, compile it into Java byte code using the Java compiler, and then run this byte code on the Java VM. In Android, things are different. You still write the Java source file, and you still compile it to Java byte code using the same Java compiler. But at that point, you recompile it once again to Dalvik byte code using the Dalvik compiler. It is this Dalvik byte code that is then executed on the Dalvik

VM. Figure 3-2 illustrates this comparison between standard Java (on the left) in Android using Dalvik (on the right).

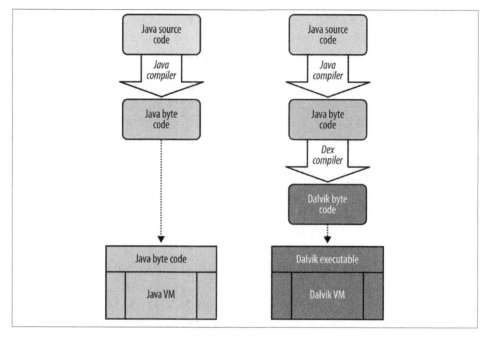

Figure 3-2. Java versus Dalvik

 It might sound like you have to do a lot more work with Android when it comes to Java. However, all these compilation steps are automated by tools such as Eclipse or Ant, and you never notice the additional steps.

You may wonder, why not compile straight from Java into the Dalvik byte code? There are a couple of good reasons for the extra steps. Back in 2005, when work on Dalvik started, the Java language was going through frequent changes, but the Java byte code was more or less set in stone. So, the Android team chose to base Dalvik on Java byte code instead of Java source code.

A side effect of this is that you can write Android applications in another language that compiles down to Java byte code. For example, you could use Scala, or Python, or Ruby to code your Android app. We're seeing some early development of apps and frameworks to support other languages and make Android development appealing to an even wider developer audience.

Another thing to keep in mind is that Android Java is a nonstandard collection of Java classes. Java typically ships in:

Java Standard Edition
Used for development on basic desktop-type applications

Java Enterprise Edition (a.k.a. J2EE or JavaEE)
Used for development of enterprise applications

Java Micro Edition (a.k.a. J2ME or JavaME)
Java for mobile applications

Android's Java set of libraries is closest to Java Standard Edition. The major difference is that Java user interface libraries (AWT and Swing) have been taken out and replaced with Android-specific user interface libraries. Android also adds quite a few new features to standard Java while supporting most of Java's standard features. So, you have most of your favorite Java libraries at your disposal, plus many new ones.

Application Framework

The application framework is a rich environment that provides numerous libraries and services to help you, the app developer, get your job done. This is the best-documented and most extensively covered part of the platform because it is this layer that empowers developers to get creative and bring fantastic applications to the market.

In the application framework layer, you will find numerous Java libraries specifically built for Android. These purpose-built Android classes live in `android.*` packages.

Yes, you also have access to most of the standard Java libraries, such as `java.lang.*`, `java.utils.*`, `java.io.*`, `java.net.*`, and so on. But some, such as `java.awt.*` and `javax.swing.*`, have been taken out. We discussed this a bit in "Android and Java" on page 37.

You will also find many services (or *managers*) that provide the ecosystem of capabilities your application can tap into, such as location, sensors, WiFi, telephony, and so on. We will talk about system services later in the book.

As you explore Android application development, most of your focus will be on this part of the stack, and you will get to use many of the application framework components.

The Android.com website provides a very good reference documentation for the entire application framework layer (*http://develop er.android.com/reference/packages.html*). While there, notice that in the top-right corner you have a search box allowing you to quickly find a reference to any Android library. On the left side of the page, you can even use the filter to filter the results based on the API level that you are developing for. For example, if you're developing your app specifically for Ice Cream Sandwich, you'd set API level to 14.

Applications

At the end of the day, we have apps. After all, the whole point of the entire set of layers of the stack below is to support the applications in providing some sort of utility to the user.

Apps can come preinstalled on the device by the carrier or manufacturer or can be downloaded by the user from one of the Android markets.

Android Application Package (APK)

An application is a single file. We call it an *Android application package*, or APK for short. An APK file has a couple of main components archived together. It is a ZIP file that you can unzip and look inside, if you're curious.

An APK consists of the following major components:

Android Manifest file
   This is the main file that provides the big picture about your app—all of its components, permissions, version, and minimum API level needed to run it, to name a few. We'll explore *AndroidManifest.xml* in much more detail in "Hello World!" on page 48.

Dalvik executable
   This is all your Java source code compiled down to a Dalvik executable. The Dalvik executable is the code that runs your application. It is located in a file called *classes.dex*.

Resources
   Resources are everything that is not code. Your application may contain a number of images and audio/video clips, as well as numerous XML files describing layouts, language packs, and so on. Collectively, these items are the resources. They are in a file called *resources.ap_* inside the APK archive as well as in subdirectories such as *drawable* for images.

Native libraries

Optionally, your application may include some native code, such as C/C++ libraries. These libraries could be packaged together with your APK file.

Signatures

Your APK file also contains a digital signature certifying that you are the author of this application. Signatures are in the *META-INF* folder. The next section describes application signing.

Application Signing

Android applications must be signed before they can be installed on a device. For development purposes, we'll be signing our example applications with a debug key—a key that you already have on your development platform. However, when you distribute your application commercially, you must sign it with your own key. The Android developer document titled "Signing Your Application" (*http://developer.android.com/ guide/publishing/app-signing.html*) has the details.

Application Distribution

One way in which Android is quite different from other platforms is the distribution of its apps. On most other platforms, such as iOS, a single vendor holds a monopoly over the distribution of applications. Android allows many different stores, or markets. Each market has its own set of policies with respect to what is allowed, how the revenue is split, and so on. As such, Android is much more of a free market space in which vendors compete for business.

In practice, this free market is very much an oligopoly, with a few big markets and many smaller boutiques.

Google Play

The biggest market currently is Google Play, also known as Play Store (formerly "Android Market"), run by Google. All the major carriers and manufacturers have it preinstalled on their devices in order to provide users with the most apps. Google knows this and uses this near-monopoly as the tool to ensure those devices adhere to Android Compatibility Test Suite, or CTS for short. We discussed CTS in "Android Compatibility" on page 4.

Other markets

In addition to Google's own market, there are many smaller boutiques. Some of them are sponsored by carriers, such at T-Mobile, Sprint, and Verizon. Others may be run by specific manufacturers, such as Cisco and its enterprise app market to support the Cisco

Cius business tablet. Additionally, enterprises are starting to roll out their own private boutiques to support their workforce, such as various US government departments.

A notable exception in this group is Amazon's App Store, which has big ambitions. It is also designed to support its Kindle Fire, an Android device that does not adhere to CTS and thus doesn't get the access to Google Play.

Side-loading apps

Applications can also be distributed via the network or via the USB cable. When you download an APK file from a website through the browser, the application represented by the APK file is installed automatically on your phone. In the development mode, we'll be using the ADB over USB to install apps on the device.

What about viruses, malware, spyware, and other bad things?

Given Android's decentralized application distribution system, it is certainly possible for an unsuspecting user to download a malicious app that consequently does bad things. For example, there have been reports of phishing attacks via fake banking apps (*http://aol.it/18kjpmu*). Android has succeeded at becoming the number one platform for vicious apps (*http://info.publicintelligence.net/DHS-FBI-AndroidThreats.pdf*).

Doesn't all this create an issue for the users? It certainly appears so. Android leaves it to the marketplace to sort itself out. Eventually, there will be stores that are more reputable and those that are less so, at least in theory. Google relies on user reports for policing its Google Play, but other markets may choose to do more proactive testing and raise the bar on what gets into the store in the first place.

Summary

In this chapter, you got a big-picture overview of what comprises the Android operating system and how its various pieces fit together. You now understand what makes Android so complete, open, and attractive to developers.

In the next chapter, we'll look at how to set up your development environment so you can get up to speed quickly. We'll also look at a simple "Hello World" application and dissect it to help you understand the various pieces of an Android application.

## CHAPTER 4

Installing and Beginning Use of Android Tools

In this chapter, you will learn how to set up your environment for Android development. We'll go beyond just listing where you can download the software, and will cover some of the best practices in getting set up. We'll look at choices for development operating systems as well as the Android tools available. You will see the good, the bad, and the ugly of the various tool and platform choices that you're about to make (or that someone else has already made for you).

By the end of this chapter, you will have your entire development environment set up. You'll be able to write a Hello World application, build it, and run it on the emulator (or a physical device, if you want).

 We use ~ to refer to your home directory. On Mac OS X, that's typically something like */Users/marko*. On Linux, it would be */home/ marko*, on Windows Vista and 7, it would be *C:\Users\marko*, and on Windows XP it would be *C:\Documents and Settings\marko*. To keep things simple and consistent, we're going to use Unix-style forward slashes and not Windows backslashes to denote file path separators.

So, if you're on Windows, just change ~ to *C:\Users* **YourUser-Name** and / to \. Other than that, everything should be pretty much the same for different operating systems, regardless of whether you use OS X, Linux, or Windows.

Installing Java Development Kit

Android development is based on Java language, tools, and libraries. So one of the first requirements is that you install Java on your machine. Before proceeding, you may want to check whether you already have Java, and whether it's an up-to-date version. To do this, open your command-line terminal:

On Windows
Click Start, choose Run, and type cmd. This should open up a command prompt window.

On Mac
Start the Terminal application located in the */Applications/Utilities/* folder.

On Linux
Open the Terminal application.

In your terminal, type **java -version** and press Enter. If the Java runtime environment is set up, you should see a version number. Make sure it is 1.6 or greater.

Next, type **javac -version** to check whether you have a Java compiler installed. You should see a version number of 1.6 or greater as well. Example 4-1 shows an example of the desired outcome.

Example 4-1. Example of Java command-line output

```
[marko:~]> java -version
java version "1.6.0_31"
Java(TM) SE Runtime Environment (build 1.6.0_31-b04-413-10M3623)
Java HotSpot(TM) 64-Bit Server VM (build 20.6-b01-413, mixed mode)
[marko:~]> javac -version
javac 1.6.0_31
```

If you pass these two tests, you can proceed to "Installing the Android SDK" on page 45. Otherwise, continue.

Mac users can install Java directly from the Software Update app. Linux users may have an automated package installation utility, depending on the Linux flavor. Windows users should install it via a download from the official Oracle site (*http://bit.ly/TEA7iC*). You want the Java Development Kit (JDK) Standard Edition (SE), version 1.6 or later.

Java comes as a Runtime Environment (JRE) and Development Kit (JDK). To program for Java, you need the JDK, which includes the Runtime Environment. The JRE on its own is good only for running existing Java code.

Java also ships in three editions: Standard Edition (JavaSE), which is your basic Java; Enterprise Edition (JavaEE, also known as J2EE), which is a bloated enterprise superset of libraries and tools; and Mobile Edition (JavaME), which is used by some mobile devices, but not by Android devices. What you need is Java SE.

Regarding versions, Android was initially based on Java version 1.5, a.k.a. Java 5. Since Gingerbread, it has been upgraded to Java 1.6, a.k.a. Java 6. So, version 1.6 is what you want. Note that as of right now, Java 1.7 or Java 7 is not fully supported by Android, nor does Android need any Java 7 features. So if possible, stay away from it. If you must use it, you can make it work for Android by setting it to behave as Java 1.6 (in Eclipse, go to Preferences → Java → Compiler and set the Compiler compliance level to 1.6 or above).

After you download Java Development Environment Standard Edition 1.6 or later for your appropriate operating system, you can usually set it up just by running the automated installation script. Repeat the command-line terminal tests discussed at the beginning of this section to make sure the installation was successful and you have the right version of Java installed.

Installing the Android SDK

The Android Software Development Kit (SDK) is necessary to develop applications for Android. The SDK comes with a set of tools as well as a platform where you can run programs and see it all work. You can download the Android SDK for your particular platform from the Android SDK Download page (*http://bit.ly/sdkandroid*).

Once you download it, unzip (or on Linux, untar) it into a folder that is easy to get to. Further examples in the book will assume your SDK is in the folder *~/android-sdk*. If it's in a different location, use that location instead of *~/android-sdk*. For example:

Windows
    C:\apps\android-sdk-windows

Linux
    */home/**YourUserName**/android-sdk-linux_86*

Mac OS X
    */Users/**YourUserName**/android-sdk-mac_86*

 For Windows users, we strongly recommend choosing directories whose names contain no spaces. This is because we'll be doing work on the command line and spaces just complicate things. Because the Windows XP home directory is in *C:\Documents and Settings*, we would recommend putting *android-sdk* in a top-level directory that you create, such as *C:\apps*.

However, on Windows Vista or 7, you can simply extract *android-sdk* into *C:\Users**YourUserName***.

Setting Up a PATH to Tools

The major tools in the Android SDK are located in two folders. Because we're going to use these tools from the command line, it is *very* helpful to add your *~/android-sdk/tools/* and your *~/android-sdk/platform-tools/* directories to your system PATH variable.

This will make it easier to access your tools without having to navigate to their specific location every single time.

Details for setting up the PATH variable depend on the platform; see step 2 of the document Installing Android SDK (*http://developer.android.com/sdk/installing.html*).

Installing Eclipse

Eclipse is an open source collection of programming tools originally created by IBM for Java. Nowadays, most developers in the Java community favor Eclipse as their Integrated Development Environment (IDE) of choice. Eclipse lives at *http://eclipse.org*.

Eclipse has a lot of time-saving features, which we'll be pointing out as we continue. Keep in mind that, although powerful, Eclipse tends to be very resource hungry, so you might want to restart it once a day if it starts running sluggishly.

Although you can do Android development with any text editor or IDE, most developers seem to be using Eclipse, and thus that's what we use in this book.

 If you choose not to use Eclipse, please refer to "Setting Up an Existing IDE" (*http://developer.android.com/sdk/installing/index.html*).

In May 2013, at Google I/O, the Android team announced it would move away from Eclipse to a new standard platform based on Gradle. Some time will pass before the tools are stable, though, so we recommend you use the existing Eclipse platform for now.

Download Eclipse (*http://www.eclipse.org/downloads/*). We recommend Eclipse IDE for Java Developers (*not* the twice-as-large Eclipse for Java EE Developers). You can install it in any directory you'd like.

Eclipse Workspace

Eclipse organizes all your work by projects. All your projects are placed in a workspace, which is a location you choose. So, your decision about where to put your workspace is significant. We recommend ~/*workspace* as a simple place for your code. On Windows, however, we recommend storing your workspace in a directory that doesn't have spaces in it (they complicate anything you might do at the command line). *C:\workspace* is a good choice for Windows users.

Setting Up Android Development Tools

After installing Eclipse and the ADK, you need to set up Android Tools for Eclipse. The instructions are as follows:

1. Start Eclipse, then select Help → Install New Software (see Figure 4-1).

2. In the Available Software dialog, click Add.

3. In the Add Site dialog that appears, enter a name for the remote site (for example, "Android Plugin") in the Name field.

4. In the Location field, enter this URL: **`https://dl-ssl.google.com/android/eclipse/`**.

5. Click OK.

6. Back in the Available Software view, you should now see Developer Tools added to the list. Select the checkbox next to Developer Tools, which will automatically select the nested tools Android DDMS and Android Development Tools. Click Next.

7. In the resulting Install Details dialog, the Android DDMS and Android Development Tools features are listed. Click Next to read and accept the license agreement and install any dependencies, then click Finish.

8. Restart Eclipse.

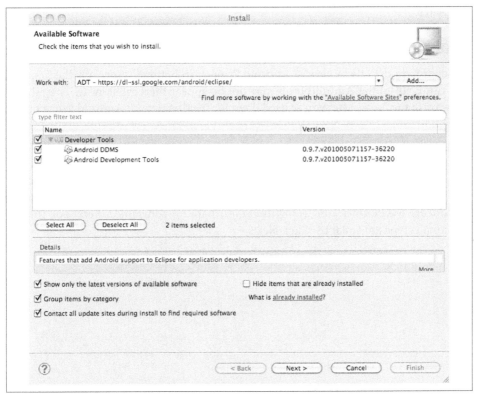

Figure 4-1. Install new software

If you have trouble downloading the plug-in, you can try using "http" in the URL instead of "https" (https is preferred for security reasons).

Eclipse is a very feature-rich development environment with many useful shortcuts and features. To help you pick up Eclipse features in the shortest time possible, we recommend watching Dan Rosen's video tutorial Introduction to Eclipse - Driving Java Productivity (*http://marakana.com/f/595*). It's only 30 minutes long and will likely save you hours as you venture into Android development.

Hello World!

To make sure everything is set up properly, we're going to write a simple Hello World program. As a matter of fact, there's not much for us to write, but a lot to understand. This is because Eclipse will create the project shell for us from some predefined templates.

Creating a New Project

In Eclipse, choose File → New → Android Project. Sometimes (especially the first time you run Eclipse) the Android tools may not be appear there right away. They should show up in the future after you've used them for the first time. If Android Project is not an option under File → New, choose Other and look for Android Project in there.

In the New Android Project dialog window (Figure 4-2), fill out the following:

- *Project name* is an Eclipse construct. Eclipse organizes everything into projects. A project name should be one word. We like to use the CamelCase (*http://en.wikipedia.org/wiki/Camel_case_(programming)*) naming convention here. For this example, type **HelloWorld**. Click Next.

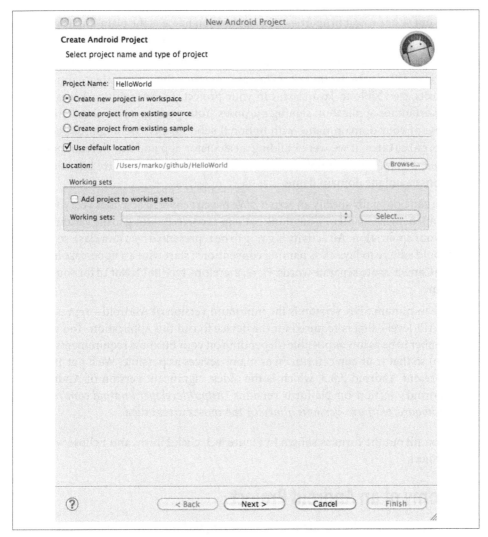

Figure 4-2. HelloWorld New Project dialog

- Choose the *build target*, which tells the build tools which version of the Android platform to build the app for. Here you should see a list of available platforms and add-ons you have installed as part of your SDK. Pick one of the newer ones, such as Android 4.x (but don't choose the targets named Google APIs—those are Google's proprietary extensions to the Android platform). For our purposes, we'll stick to Android Open Source versions of the Android platform. Click Next.

- Fill out your project properties. The application name is the plain English name of your application. Enter something like `Hello, World!!!`.

- The package name is a Java construct. In Java, all source code is organized into packages. Packages are important because, among other things, they control which objects are visible to Java classes in your project. In Android, packages are also important for application signing purposes. Your package name should be the reverse of your domain name with optional subdomains. We might use `com.exam ple.calculator` if we were building a calculator app and our domain name was *example.com*. This example uses `com.marakana.android.hello`, but you should choose a unique domain name.

- You can optionally specify an activity. We haven't covered activities yet (you'll learn about them in Chapter 7), but think of them as corresponding to the various screens in your application. An activity is going to be represented by a Java class, so its name should adhere to Java class naming conventions: start with an uppercase letter and use CamelCase to separate words. Here, therefore, type `HelloWorld` for your activity name.

- The minimum SDK version is the minimum version of Android—represented by its API level—that is required for the device to run this application. You want this number to be as low as possible (depending on your business requirements for your app) so that your app can run on as many devices as possible. We'll put 10 here to represent Android 2.3.3, which is the oldest significant version of Android. See Android's section on platform versions (*http://developer.android.com/resources/ dashboard/platform-versions.html*) for the most current data.

After you fill out the form as shown in Figure 4-3, click Finish, and Eclipse will create your project.

Anatomy of an Android Project

You have created a Hello World project, but so far you haven't written much. Yet Eclipse has created a whole new project populated with numerous files. The directory structure with all these files can be viewed in Eclipse's Package Explorer window (Figure 4-4). These boilerplate files represent a structure of a typical Android project. In this section, we'll review its major parts.

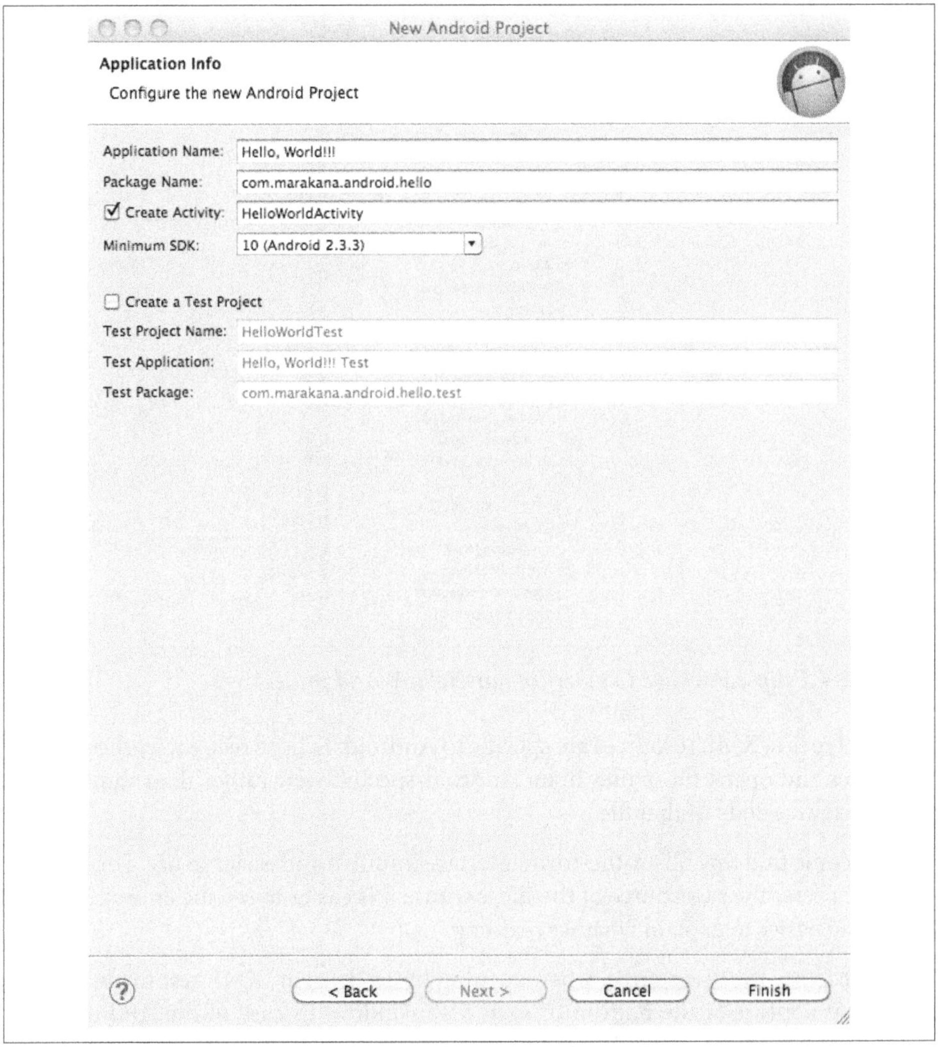

Figure 4-3. HelloWorld New Project dialog: Application Info

Android Manifest File

The manifest file glues everything together. It explains what the application consists of, what all its main building blocks are, what permissions it requires, and so on. To look at its contents, double-click the *AndroidManifest.xml* file in Eclipse's Package Explorer window.

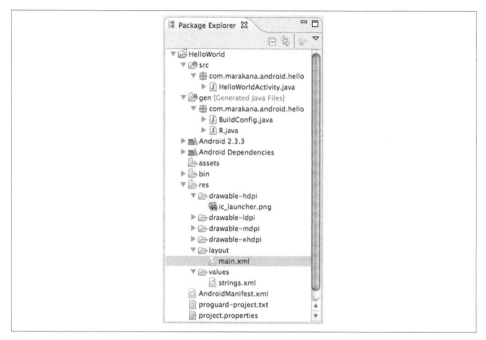

Figure 4-4. Eclipse Package Explorer of our Hello World project

Certain types of XML resources are specific to Android. Eclipse recognizes these special resources and opens those files in an Android-specific view rather than showing you the raw source code of that file.

Some people find this fill-in-the-form interface intuitive and easier to use. For the most part, we prefer the raw source of the file, because it is easier to see the entire content at once, and easier to explain what we're doing.

In order to see the raw source, when opening those "known" XML resources, look for a tab that appears at the bottom of each file's window. In case of the *AndroidMani-fest.xml* file, the bottom tab bar contains Manifest, Application, Permissions, Instrumentation, and AndroidManifest.xml tabs. These are different views of the same source file, and it's the Eclipse attempt at providing hand-holding in editing the file itself. In order to get to the actual source of the file, choose the AndroidManifest.xml tab on the bottom right, as shown in Figure 4-5.

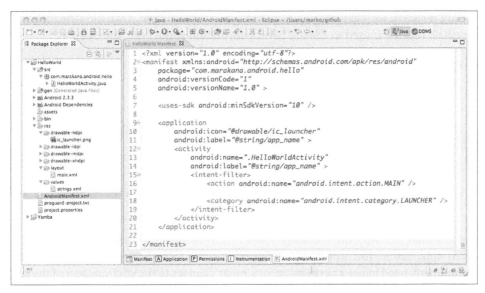

Figure 4-5. Eclipse—Java perspective

Example 4-2 contains the source code of this file.

Example 4-2. AndroidManifest.xml

```xml
<?xml version="1.0" encoding="utf-8"?>
<manifest xmlns:android="http://schemas.android.com/apk/res/android"
    package="com.marakana.android.hello"
    android:versionCode="1"
    android:versionName="1.0" >

    <uses-sdk android:minSdkVersion="10" />

    <application
        android:icon="@drawable/ic_launcher"
        android:label="@string/app_name" >
        <activity
            android:name=".HelloWorldActivity"
            android:label="@string/app_name" >
            <intent-filter>
                <action android:name="android.intent.action.MAIN" />

                <category android:name="android.intent.category.LAUNCHER" />
            </intent-filter>
        </activity>
    </application>

</manifest>
```

Going forward, we'll assume you have basic understanding of XML file structure. In a nutshell, this file is used to declare basic information about your app, such as:

- The package under which it's registered
- The version of the app, both code and name
- The Android SDK that it is targeting and requiring in order to run
- The permissions that it uses in order to run (the user gets asked to grant them at install time)
- Custom permissions that it declares and may require from other components
- The application and all its main building blocks: activities, services, providers, and receivers

An Android Manifest file is usually about one to two screens in length in a medium-sized application. You will find yourself editing this file whenever you create a new component, need to use a permission, or similar changes to the environment of the app.

Next, we'll look at some of the resources that are created for our Hello, World project.

String Resources

Moving up the list of significant files created from the boilerplate template for a new Android project, the first resource file, *strings.xml*, is located in the *res/values/* folder of your project.

This is another XML file that contains all the text that your application uses: for example, the names of buttons, labels, default text, and similar types of strings. This is the best practice for separating the concerns of various files, even if they are XML files. In other words, *layout.xml* is responsible for the layout of widgets, but *strings.xml* is responsible for their textual content (see Example 4-3).

Just as with the Manifest file, you can view this file with the so-called *Android Localization Files Editor*, or you look at the raw source code by clicking the *values/ strings.xml* tab on the bottom right. Again, we prefer the source code view so you can see the entire file in its entirety.

Example 4-3. The res/values/strings.xml file

```
<?xml version="1.0" encoding="utf-8" standalone="no"?>
<resources>
    <string name="hello">Hello World, HelloWorldActivity!</string>
    <string name="app_name">Hello, World!!!</string>
</resources>
```

The strings resource file is simply a set of name-value pairs, where the name is the name of a string resource and the value is its actual text. By referring to strings by their made-up names, we can later change the actual value without changing any of our Java code. This is an important feature, because any changes to code would require additional testing of the application before releasing it to the user base.

Additionally, by using a name rather than the actual value, we can provide multiple sets of values for the same string resource. In other words, the same text could be provided in multiple languages. We'll explore this feature in more detail in "Alternative Resources" on page 124. For now, it's good to know you can have multiple versions of the *strings.xml* file for each language you want your app to be supported in. The Android OS will automatically determine the most appropriate "language pack" to use, or will just revert to default one in case there's nothing that matches user's preferred locale.

Layout XML Code

Next up is the *res/layout* folder, containing the *main.xml* file. This is an XML file declaring the layout of our screen.

As before, there are two ways to look at this file: graphically and as the raw XML. When working with layouts, graphical view can be a very valuable tool. There you can drag and drop various widgets on the screen, move them around, and right-click particular component to set one of numerous properties. The graphical layout for our basic *main.xml* file is shown in Figure 4-6.

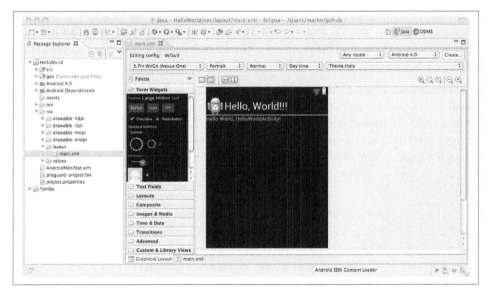

Figure 4-6. Graphical layout view

As before, you can click the bottom-right tab to see the actual source code of this file, as shown in Example 4-4.

Example 4-4. The res/layout/main.xml file

```xml
<?xml version="1.0" encoding="utf-8"?>
<LinearLayout xmlns:android="http://schemas.android.com/apk/res/android"
    android:layout_width="fill_parent"
    android:layout_height="fill_parent"
    android:orientation="vertical" >

    <TextView
        android:layout_width="fill_parent"
        android:layout_height="wrap_content"
        android:text="@string/hello" />

</LinearLayout>
```

You can think of this XML file as an HTML page. It declares the components of a single screen in our Android app. Just as tools such as Adobe's Dreamweaver allow you to create a web page in What-You-See-Is-What-You-Get (WYSIWYG) mode as well as in raw HTML view, Eclipse lets you use both graphical layout view as well as edit the raw XML source code.

Drawable Resources

Next, the list of files in our Hello World project in the Eclipse Package Explorer is a set of *drawable* folders. They are also resources, but this time they are images. If you expand all these various drawable folders, you'll notice that they all contain the *ic_launcher.png* file. This is the icon that appears in the Launcher app that we click to launch our application.

The reason that we have three or four separate folders is that each one is for a different density of the screen. So *drawable-hdpi* will be used when the app is running on a device with a high-density screen (240 dots per inch), *drawable-mdpi* is used on medium screens (160 dpi), and *drawable-ldpi* is used on low-density screens (80 dpi). And although it appears they all contain the same image, the images are actually different sizes, so that on various screens they appear correct, taking into account number of pixels available per inch.

This is an example of how alternative resources work. The same functionality is available to all other resources as well. We briefly discussed this in "String Resources" on page 54 and will cover it in more detail in "Alternative Resources" on page 124.

The R File

We are finally making it up to the world of Java files. The *gen* folder is where all the auto-generated files are located. For now, *R.java* is the only significant one, with

BuildConfig.java being introduced in later versions of the Ice Cream Sandwich release of tools (ADK R18).

The R file is the glue between the world of Java and the world of resources (see Example 4-5). It is an automatically generated file, and as such, you never modify it. It is recreated every time you change anything in the *res* directory; for example, when you add an image or modify an XML file.

You don't need to look at this file much in the future. We will use the data in it quite a bit, but we'll use Eclipse to help us refer to values stored in this file.

Example 4-5. gen/com/marakana/R.java

```
/* AUTO-GENERATED FILE.  DO NOT MODIFY.
 *
 * This class was automatically generated by the
 * aapt tool from the resource data it found.  It
 * should not be modified by hand.
 */

package com.marakana;

public final class R {
    public static final class attr {
    }
    public static final class drawable {
        public static final int icon=0x7f020000;
    }
    public static final class layout {
        public static final int main=0x7f030000;
    }
    public static final class string {
        public static final int app_name=0x7f040001;
        public static final int hello=0x7f040000;
    }
}
```

This file represents a set of various pointers that allow Java to locate app resources at runtime. The entire file, including the obscure values, is automatically generated by an SDK tool called *aapt* and has little value to humans. We'll see later on how we use this file, starting with the Java source code in the following section.

Java Source Code

At the top, we finally have our Java code, which drives everything. It represents the core logic behind our application. As such, it is the main starting point of our app execution.

Example 4-6 contains the source of our Java code representing the Hello World activity; in other words, the logic behind the single screen that we have in our app.

Example 4-6. HelloWorld.java

```java
package com.marakana.android.hello;

import android.app.Activity;
import android.os.Bundle;

public class HelloWorldActivity extends Activity {
    /** Called when the activity is first created. */
    @Override
    public void onCreate(Bundle savedInstanceState) {
        super.onCreate(savedInstanceState);
        setContentView(R.layout.main);
    }
}
```

What goes on in this file will become much more apparent later on. For now, just notice that we're subclassing the `Activity` class, which we included from the Android framework.

So now that we've covered all the significant files that make up the Hello World project, we're ready to test-drive this application.

Building the Project

Thus far, we haven't talked about how to build our project. In a traditional software project, at some point you have to compile it, link it with other libraries, and finally create a shippable product that users can install and run.

Eclipse actually takes care of much of the build process for you. This is because, by default, the Build Automatically feature is enabled under the File → Project menu. If it's not, you may want to enable it, because it is usually a very useful feature. When Build Automatically is on, every time you change any file, or anything in your project, Eclipse automatically rebuilds the application. The by-products of this build are under the *bin* folder in your project. If you look under the bin folder, you will find the following:

classes folder
> This is where your Java source code is compiled into Java byte code, or *.class* files.

classes.dex file
> Once your Java is compiled, it is recompiled once again into the Dalvik byte code, and *classes.dex* is the archive containing all those recompiled Dalvik classes. We talked about this process in "Android and Java" on page 37.

res folder
> This is where any binary resources are copied over, such as images, movies, and audio clips.

---

resources.ap file
> This file is an archive of all the XML resource files encoded in an efficient and easy-to-parse format.

HelloWorld.apk
> This is our final shippable product. The APK (Android Package) file represents our application in its entirety. It is the archive containing all the Dalvik executable code and all the resources, as well as the Android Manifest file that describes the application's metainformation.

Now that we have the final shippable product in the form of our *HelloWorld.apk* file, we're ready to execute it on an Android device.

Android Emulator

The Emulator is a tool that ships with Android SDK. It allows you to run any number of Android Virtual Devices (AVDs) right on your computer without needing a real Android device for development purposes. The Emulator not only opens Android app development to way more programmers, but it also makes the whole program-deploy-test cycle shorter, making the Android development more enjoyable.

 A simulator and an emulator sound very similar, but are fundamentally different. To emulate means to imitate the machine executing the binary code. So, an emulator is sort of like a virtual machine. A simulator merely simulates the behavior of the code at a higher level. Android SDK ships with a true emulator, based on QEMU (*http://wiki.qemu.org/Main_Page*).

An Emulator Versus a Physical Phone

For the most part, running your application on the emulator is identical to running it on a physical phone. That is because the emulator is an actual code emulator, meaning it runs the same code base as the actual device, all the way down to the machine layer.

There are some notable exceptions, mostly things that are just hard to virtualize, such as sensors. Other hardware-related features, such as telephony and location services, can be simulated in the emulator.

Unless otherwise noted, we're going to be agnostic to the fact that your code may be running on a real devices versus an emulated device.

Until the Honeycomb version of Android, the Emulator was reasonably fast, and for the most part we'd prefer testing our code on the emulated device versus the real one. With Honeycomb, the Emulator became extremely slow and painful to use. This mostly has to do with lack of multicore support and lack of GPU support that real devices enjoy and that are missing from the Emulator. The Android team realized this shortcoming and, since later versions of Ice Cream Sandwich (ADT R18), th Emulator supports GPU, making it substantially faster. It is, once again, enjoyable to test development code on it.

To use the emulator, we'll have to create an Android Virtual Device (AVD). The easiest way to do that is to start Android Virtual Device tool in Eclipse.

To create a new AVD, start the Android Virtual Device Manager. You can start this tool from Eclipse by clicking the ⊞ icon or by choosing Window → AVD Manager in the Eclipse menu bar. You should get a dialog window that looks similar to Figure 4-7.

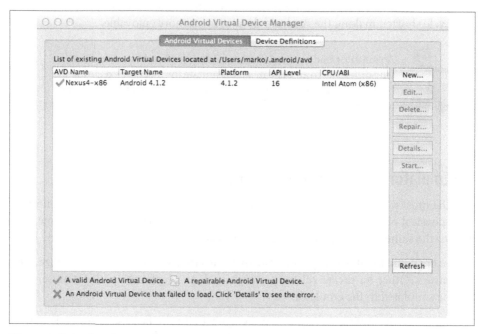

Figure 4-7. Android Virtual Device Manager

From within the Android Virtual Device Manager window, choosing New pops up a Create New AVD dialog window (see Figure 4-8). In this dialog, specify the parameters for your new AVD. The name can be any name you choose. The target designates which version of Android you want installed on this particular AVD. The list of possible targets is based on platforms and add-ons that you have installed into your SDK. If you don't have any targets, go back to the Android SDK and AVD Manager window and choose the "Available packages" tab to install at least one platform, for example, Android 4.1 (API level 16).

Each AVD can have an SD card. You can just specify a number here for your built-in card, in megabytes. The skin is the look and feel of your device as well as its form factor. The Hardware option lets you fine-tune what this AVD does and doesn't support.

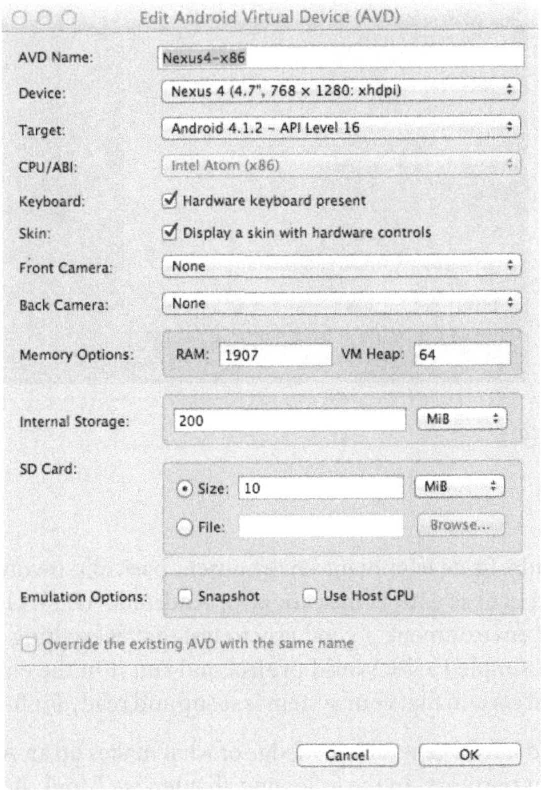

Figure 4-8. Android Virtual Device Manager: Create New AVD dialog

Once you are done with this dialog, you will have a new Android Virtual Device in your list. Go ahead and start it by clicking Start and then Launch, and an emulator will pop up (see Figure 4-9).

Figure 4-9. Emulator

Summary

Setting up the Android development environment basically involves setting up Java, Android SDK, and Eclipse with Eclipse tools for Android (ADT). Once you have set up your development environment, a good way to test that everything is working is to use Eclipse to create a simple Hello, World project and run it in the emulator. If that runs fine, you are almost certain that your system is set up and ready for further development.

By now, you should have a general knowledge of what makes up an Android application and what the major parts are. In the following chapter, we'll look at how to architect an Android app using Android's main building blocks.

## CHAPTER 5

Main Building Blocks

In this chapter, you will learn Android's capabilities by looking at the big features it offers. We'll give you a high-level overview of what activities are, how intents work, when and how to use services, how to use broadcast receivers and content providers to make your app scale, and much more.

By "main building blocks," we refer to the pieces of an application that Android offers you to put together into an Android app. When you start thinking about your application, it is good to take a top-down approach. For instance, most programmers design applications in terms of screens, features, and the interactions between them. You start with a conceptual drawing, something that you can represent in terms of "lines and circles." This approach to application development helps you see the big picture—how the components fit together and how it all makes sense.

By the end of this chapter, you will understand the main Android components for building applications. You should conceptually know when you'd use what component. You will also see how these components relate to a real-world application.

A Real-World Example

In this book, we're going to build an app to use Twitter. We know that the user should be able to post status updates. We also know the user should be able to see what her friends are up to. Those are basic features. Beyond that, the user should also be able to update her username and password for the online account. So now we know we should have the following three screens: a screen for users to update their own status, a screen to see the status timelines of their friends, and a screen to set the preferences for the app.

Next, we would like this app to work quickly regardless of the network connection or lack thereof. To achieve that, the app has to pull the data from the cloud when it's online and cache the data locally. That will require a service that runs in the background as well as a database.

We also know that we'd like this background service to be started when the device is initially turned on, so by the time the user first uses the app, there's already up-to-date information in the local cache.

Finally, we will want to display the latest status from the friends' timelines on the home screen, as an Android Widget.

These are some straightforward requirements. Android building blocks make it easy to break them down into conceptual units so that you can work on them independently, and then easily put them together into a complete package.

As you'll see later, these building blocks make up an application. Essentially, an app is not much more than a loose collection of activities, services, providers, and receivers.

Activities

An activity is usually a single screen that the user sees on the device at one time. An application typically has multiple activities, and the user flips back and forth among them. As such, activities are the most visible part of your application.

We usually use a website as an analogy for activities. Just as a website consists of multiple pages, so does an Android application consist of multiple activities. Just as a website has a "home page," an Android app has a "main" activity, usually the one that is shown first when you launch the application. And just as a website has to provide some sort of navigation among various pages, an Android app should do the same.

On the Web, you can jump from a page on one website to a page on another. Similarly, in Android, you could be looking at an activity of one application, but shortly afterward you could start another activity in a completely separate application. For example, if you are in your Contacts app and you choose to text a friend, you'd be launching the activity to compose a text message in the Messaging application.

Activity Life Cycle

Launching an activity can be quite expensive. It may involve creating a new Linux process, allocating memory for all the UI objects, inflating all the objects from XML layouts, and setting up the whole screen. Because the operating system is doing a lot of work to launch an activity, it would be a waste to just toss it out once the user leaves that screen. To avoid this waste, the activity life cycle is managed by the Activity Manager, a service that runs inside the Android Framework layer of the stack.

The Activity Manager is responsible for creating, destroying, and managing activities. For example, when the user starts an application for the first time, the Activity Manager will create its activity and put it onto the screen. Later, when the user switches screens, the Activity Manager will move that previous activity to a holding place. This way, if the user wants to go back to an older activity, it can be started more quickly. Older activities that the user hasn't used in a while will be destroyed in order to free more space for the currently active one. This mechanism is designed to help improve the speed of the user interface and thus improve the overall user experience.

Programming for Android is conceptually different from programming for some other environments. In Android, you find yourself responding to certain changes in the state of your application rather than driving that change yourself. It is a managed, container-based environment similar to programming for Java applets or servlets. So, when it comes to an activity life cycle, you don't get to say what state the activity is in, but you have plenty of opportunity to say what happens during the transitions from state to state. Figure 5-1 shows the states that an activity can go through. The following sections describe how to handle each state.

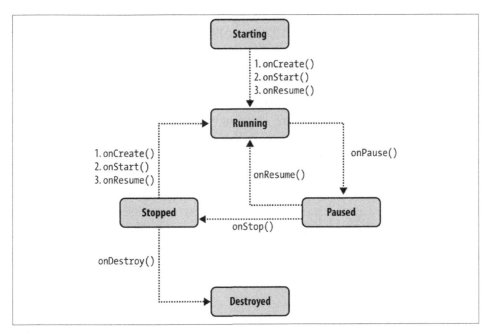

Figure 5-1. Activity life cycle

Starting state

When an activity doesn't exist in memory, it is in a *starting state*. As it starts, the activity invokes a set of callback methods that you as a developer have an opportunity to fill out. These callbacks include onCreate(), onStart(), and onResume(). Eventually, the activity will be in a *running state*, which means that it will be fully displayed on the screen, in focus, waiting for user to interact with it.

Keep in mind that this transition from starting state to running state is one of the most expensive operations the application will perform in terms of computing time, and this also directly affects the battery life of the device. This is the exact reason we don't automatically destroy activities that are no longer shown. The user might want to come back to them, so the operating system, via Activity Manager, keeps them around for some time.

Running state

Only one activity on a device can be in a running state: it's the one that is currently on the screen and interacting with the user. We also say this activity is *in focus*, meaning that all user interactions—such as typing, touching the screen, and clicking buttons—are handled by this one activity.

The running activity has priority in terms of getting the memory and resources it needs to run as quickly as possible. This is because Android wants to make sure the running activity is zippy and responsive to the user.

Paused state

When an activity is not in focus (i.e., not interacting with the user) but still visible on the screen, we say it's in a *paused state*. This is not a typical scenario, because the device's screen is usually small, and an activity is either taking up the whole screen or none at all. We often see this case with dialog boxes that come up in front of an activity, causing it to become paused. All activities go through a paused state en route to being stopped.

Paused activities still have high priority in terms of getting memory and other resources. This is because they are visible and cannot be removed from the screen without making it look very strange to the user. The Activity Manager calls onPause() when putting your application into the paused state, but we don't use that hook to perform any activities in this book.

Stopped state

When an activity is not visible, but still in memory, we say it's in a *stopped state*. A stopped activity could be brought back to the front to become a *running* activity again. Or, it could be destroyed and removed from memory, which is an operating system choice beyond your control.

The system keeps activities around in a stopped state because it is likely that the user will still want to get back to those activities some time soon, and restarting a stopped activity is far cheaper than starting an activity from scratch. That is because the Activity

Manager already has all the objects loaded in memory and simply has to bring them all up to the foreground.

Stopped activities can be removed from memory at any point. It is up to Activity Manager's discretion to do so. The Activity Manager calls onStop() when putting your application into this state, so it is wise in this method to do anything you need in order to save the state of your app, such as writing data to disk or a database.

Destroyed state

A destroyed activity is no longer in memory. The Activity Manager decided that this activity is no longer needed and has removed it. Before the activity is destroyed, it can perform certain actions, such as save any unsaved information. However, there's no guarantee that your activity will be destroyed from the *destroyed state*. It is possible for a stopped activity to be destroyed as well. For that reason, it is better to do important work, such as saving unsaved data, in the onStop() rather than the onDestroy() callback.

The fact that an activity is in a running state doesn't mean it's doing much. It could be just sitting there and waiting for user input. Similarly, an activity in a stopped state is not necessarily doing nothing. The state names mostly refer to how active the activity is with respect to user input; in other words, whether an activity is visible, in focus, or not visible at all.

Why is the process of managing the activity life cycle so complex? On a typical desktop PC, when you are done with an application, such as Microsoft Word or Excel, you close it. Essentially, what you as a user are doing is memory managing your PC. Android's team felt that users shouldn't have to manage memory and have delegated that responsibility to the Activity Manager. Sure, humans would likely do a better job of quitting apps they no longer need, but the automatic way is good enough and makes the overall experience better for the user. We sometimes compare Activity Manager to automatic transmission on a car, or garbage collection in languages such as Java. Yes, humans do a better job with switching manual gears or allocating and freeing memory manually, but that's just extra work that, in today's day and age, machines do well enough.

Because the user interface is a big part of most Android apps, we'll explore how to create activities in detail in Chapter 7.

There are currently two ways to create activity user interfaces: the standard and older activity views and the newer fragments. Fragments were introduced in Android version 3.0 as a means to simplify the handling of different screen sizes and devices. We will explore this in Chapter 8.

Intents

Intents are messages that are sent among the major building blocks. They trigger an activity to start up, tell a service to start, stop, or bind to, or are simply broadcasts. Intents are asynchronous, meaning the code that sends them doesn't have to wait for them to be completed. To use our analogy with a website, intents would be the links connecting various pages together. Just like a web link, an intent can be internal to a single app or, just as easily, connect to components in other apps. And just like links, intents can carry small amounts of primitive data with them.

An intent could be *explicit* or *implicit*. In an explicit intent, the sender clearly spells out which specific component should be on the receiving end. In an implicit intent, the sender specifies just the type of receiver. For example, your activity could send an intent saying it simply wants someone to open up a web page. In that case, any application that is capable of opening a web page could "compete" to complete this action.

When you have competing applications, the system will ask you which one you'd like to use to complete a given action. You can also set an app as the default. This mechanism works very similarly to your desktop environment, for example, when you downloaded Firefox or Chrome to replace your default Internet Explorer or Safari web browsers.

This type of messaging allows the user to replace any app on the system with a custom one. For example, you might want to download a different SMS application or another browser to replace your existing ones. Figure 5-2 shows how intents can be used to "jump" between various activities, in the same application or in another app altogether.

You will learn about how to create and use intents in "The Action Bar" on page 148 when we talk about the Action Bar in Android.

Services

Services run in the background and don't have any user interface components. They can perform the same actions as activities, but without any user interface. Services are useful for actions that you want to perform for a while, regardless of what is on the screen. For example, you might want your music player to play music even as you are flipping between other applications.

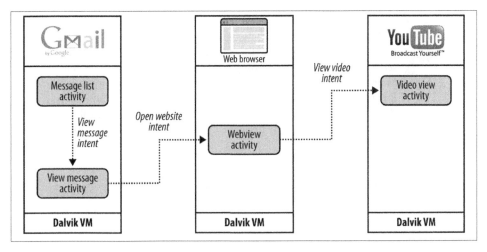

Figure 5-2. Intents

 Don't confuse the Android services that are part of an Android app with native Linux services, servers, or daemons, which are much lower-level components of the operating system.

Services have a much simpler life cycle than activities (see Figure 5-3). You either start a service or stop it. Also, the service life cycle is more or less controlled by the developer, and not so much by the system. Consequently, developers have to be mindful to run services so that they don't consume shared resources unnecessarily, such as the CPU and battery.

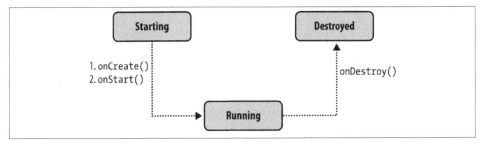

Figure 5-3. Service life cycle

Just because a service runs in the background doesn't necessarily mean it runs on a separate thread. By default, services and activities run on the same main application thread, often called the UI thread. If a service is doing some processing that takes a while to complete (such as performing network calls), you would typically invoke a separate thread to run it. Otherwise, your user interface will run noticeably slower. As of the Honeycomb release of Android, you are actually not even allowed to perform network or other potentially long operations on the UI thread.

We'll look at how to create a new service in Chapter 10.

Content Providers

Content providers are interfaces for sharing data between applications. By default, Android runs each application in its own sandbox so that all data that belongs to an application is totally isolated from other applications on the system. Although small amounts of data can be passed between applications via intents, content providers are much better suited for sharing persistent data between possibly large datasets. As such, the content provider API nicely adheres to the CRUD principle (*http://en.wikipedia.org/wiki/Create,_read,_update_and_delete*). Figure 5-4 illustrates how the content provider's CRUD interface pierces the application boundaries and allows other apps to connect to it to share data. The methods that content providers use to implement the four critical operations are:

Operation	Method
Create	insert()
Read	query()
Update	update()
Delete	delete()

The Android system uses this mechanism all the time. For example:

- The Contacts Provider exposes all user contact data to various applications.
- The Settings Provider exposes system settings to various applications, including the built-in Settings application.
- The Media Store is responsible for storing and sharing various media, such as photos and music, across various applications.

Figure 5-5 illustrates how the Contacts app uses the Contacts Provider, a totally separate application, to retrieve data about users' contacts. The Contacts app itself doesn't have any contacts data, and the Contacts Provider doesn't have any user interface.

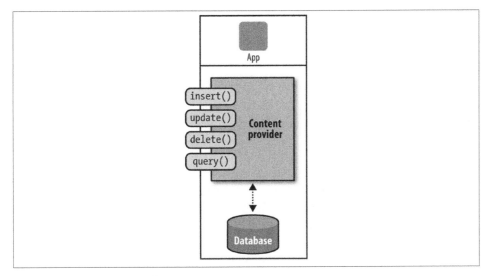

Figure 5-4. Content provider

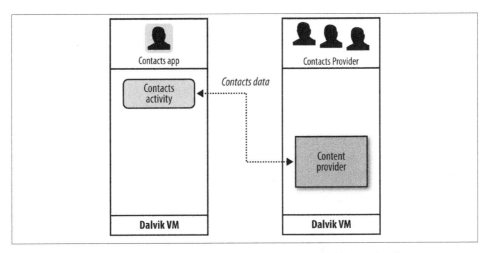

Figure 5-5. Contacts application using the Contacts Provider to get the data

This separation of data storage and the actual user interface application offers greater flexibility to mash up various parts of the system. For example, a user could install an alternative address book application that uses the same data as the default Contacts app. Or he could install widgets on the home screen that allow for easy changes in the System Settings, such as turning on or off the WiFi, Bluetooth, or GPS features. Many phone manufacturers take advantage of content providers to add their own applications on top of standard Android to improve the overall user experience, such as HTC Sense (*http://en.wikipedia.org/wiki/HTC_Sense*).

Content providers are relatively simple interfaces. The `insert()`, `update()`, `delete()`, and `query()` methods look a lot like standard database methods, so it is relatively easy to implement a content provider as a proxy to the database.

Although content providers are relatively easy to use, they are somewhat tricky to implement properly. We'll explore how to create a new content provider in Chapter 11.

Broadcast Receivers

Broadcast receivers are Android's implementation of a system-wide publish/subscribe mechanism (*http://en.wikipedia.org/wiki/Publish/subscribe*), or more precisely, an Observer pattern (*http://en.wikipedia.org/wiki/Observer_pattern*). The receiver is simply dormant code that gets activated by the occurrence of an event to which the receiver is subscribed. The "event" takes the form of an intent.

The system itself broadcasts events all the time. For example, when an SMS arrives, a call comes in, the battery runs low, or the system completes booting up, all those events are broadcast, and any number of receivers could be triggered by them.

Broadcast receivers themselves do not have any visual representation, nor are they actively running in memory. But when triggered, they get to execute some code, such as starting an activity, a service, or something else.

In our sample app mentioned earlier, we want to trigger the update of the local data from the cloud every once in a while. To do that, we can set up an alarm that fires a broadcast intent on some interval, and that intent triggers our receiver to start the refresh service. So we have an intent triggering a receiver via one type of intent that later starts a service via another type of intent. This chaining of intents is common in Android and provides for a loosely coupled architecture. We'll talk about that in the next section.

You will learn more about broadcast receivers and how to implement them in Chapter 13.

Application Context

So far you have seen activities, services, content providers, and broadcast receivers. Together, they make up an application. They are the basic building blocks, loosely coupled, of an Android app.

Think of an application in Android as a container to hold together your blocks. Your activities, services, providers, and receivers do the actual work. The container that they work in is the shared Linux process with common Dalvik VM, private filesystem, shared resources, and similar things.

To use our website analogy, an app would be the website domain. Users never really go to Amazon.com (the domain), but rather visit a web page (which you could compare to an Android activity) within that domain, or consume a web service (an Android service). So web pages and web services are building blocks for a website, and the website itself is just a container to hold them all together under one roof. This is very similar to what an Android application does for its components.

Application context refers to the application environment and the process within which all its components are running (see Figure 5-6). It allows applications to share the data and resources between various building blocks. It has its own Linux user ID and its own Linux process, with a dedicated Dalvik virtual machine, a dedicated filesystem for storing application files, and so on.

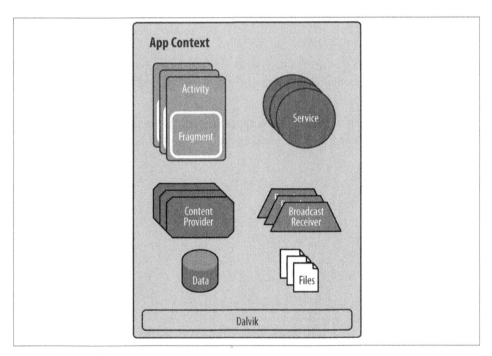

Figure 5-6. Application context

The application context is uniquely identified on a device based on the package name of that application. For example, `com.marakana.android.yamba` would be a package name of our app. There cannot be another app with the same package name (unless it comes from us, and we want to use shared user IDs).

An application context gets created whenever the first component of this application starts up, regardless of whether that component is an activity, a service, or something else. The application context lives as long as your application is alive. As such, it is independent of the activity's life cycle. So as long as any of the activities, services, providers, or receivers are alive, your application context is around to hold them. Once the Activity Manager terminates all other building blocks of your app, it also gets rid of the application context because it's no longer needed.

You can easily obtain a reference to the context by calling `Context.getApplication Context()` or `Activity.getApplication()`. Keep in mind that activities and services are already subclasses of the context, and therefore inherit all its methods.

 Activities and services are also subclasses of the `Context` class, which is different from the application context we're talking about here.

Summary

In this chapter, you learned about some of the most important Android application components.

In the next chapter, we'll outline a Yamba application as an example of how all these bits and pieces come together to form a working Android app.

## CHAPTER 6

Yamba Project Overview

The best way to learn is by example. After working with thousands of new Android developers and using various example applications to explain some of the unique concepts that this platform has to offer, we have concluded that the best example should meet the following criteria:

Be comprehensive
> A great example app should demonstrate most of the unique aspects of building apps for Android. It should cover all main building blocks and their typical usage.

Provide motivation
> There should be a good reason to use a specific feature. A great example app will defend the most effective design that also relies on Android building blocks.

Be familiar
> The example application should be simple to understand. We want to focus on software design and implementation, and not on explaining user features.

The Yamba Application

The application we picked for this book is a Twitter-like application. We call it Yamba, which stands for *Yet Another Micro Blogging App*. Yamba lets a user connect to a cloud web service, pull down friends' statuses, and update his own status.

Yamba covers most of the main Android building blocks in a natural way. As such, it's a great sample application to illustrate both how various components work individually and how they fit together. Services such as Twitter are more or less familiar to most people, so the features of the application do not require much explanation.

Yamba Versus Twitter.com

Yamba will not work with Twitter.com, at least not out of the box. We are using the same Twitter API (*https://dev.twitter.com/*). However, Twitter.com has recently changed to OAuth (*http://oauth.net/*) for its authentication and no longer supports simple (username and password) login. Although we could implement OAuth in Yamba, we felt that doing so dramatically changes our learning objectives and the flow of the material. OAuth would take us on a tangent that is not primary to learning Android development philosophy.

Additionally, when Marko runs Android training courses, expecting students to have Twitter accounts (try that in China!) and to experiment live with them on an app still in development, is not practical. Plus, we often run into Twitter's *rate limit* designed to prevent spam and DoS attacks. Sometimes our test code may seem to be doing just that.

So we have set up our own instance of a cloud service adhering to same Twitter API as the real Twitter.com. Our site is at *http://yamba.marakana.com/*. It doesn't enforce rate limits, and creating an account is easy. As such, it is designed for learning purposes and testing of your app. You, too, can roll out your own Twitter-like service using the Status.net (*http://status.net/open-source*) open source platform.

The following figures show what a finished product could look like. Figure 6-1 shows how Yamba displays a list of status messages from your friends. Figure 6-2 shows the initial Yamba screen, and Figure 6-3 shows the user preferences.

Figure 6-1. List of status messages from other people, called a timeline

Figure 6-2. Screen where the user can enter a status message

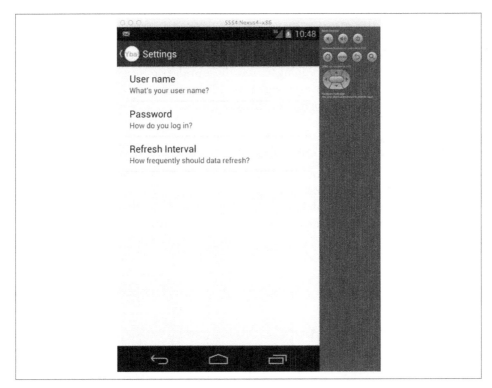

Figure 6-3. User settings

Design Philosophy

We're going to adopt a certain design philosophy in tackling this project. This philosophy will help guide us in our development and serve as a north star when in doubt about what to do next. Spelling out the design philosophy here should also help eliminate some confusion in the process we're following:

Small increments
> The Yamba application will start out small and will constantly grow in functionality and complexity. Initially, the app will not do much, but it will grow organically one step at a time. Along the way, we'll explain each step so that you're expanding your skills as you go.

Always whole and complete
> The application must always work. In other words, we'll add new features in small, self-contained chunks and pull them back into the main project so that you can see how it fits together as a whole.

Refactoring code

Once in a while, we'll have to take a step back and refactor the application to remove duplicate code and optimize the design. The goal is to reuse the code and not reinvent the wheel. But we are going to cross those bridges as we get to them, providing the motivation for refactoring along the way. This process will teach you about some general software development best practices as well.

Project Design

If you remember from Chapter 5, an Android application is a loose collection of activities, services, content providers, and broadcast receivers. These are the components from which we put together an application. Figure 6-4 shows the design of the entire Yamba application, which incorporates most of the main Android building blocks.

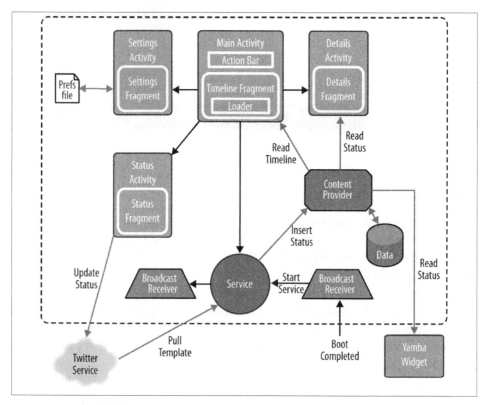

Figure 6-4. Yamba design diagram

To help understand the diagrams as we keep moving through the the design, Figure 6-5 provides a quick legend of the design language that we have developed specifically for purposes of illustrating how Yamba comes together.

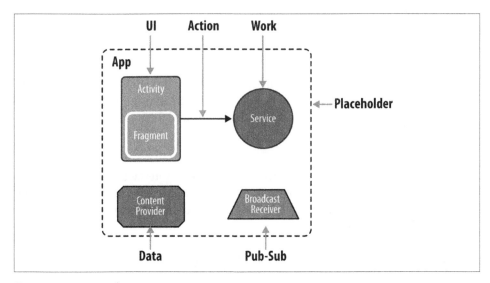

Figure 6-5. Design language

Part 1: Android User Interface

This part, covered in Chapters 7 and 8, will focus on developing the first component of the Yamba application: the Status Update screen. Our tasks are building an activity, networking and multithreading, and debugging:

Building an activity
> We are going to start by introducing the Android user interface (UI) model. In its UI, Android is quite different from some other paradigms that you might be familiar with. This is done with a dual approach to UI using both Java and XML.

> In this chapter, you will learn how to develop the user interface, as shown in Figure 6-2, where the user updates her status. Through this process, you will use XML and Java to put together a working UI. You will learn about views and layouts, units in Android, how to work with images, and how to make the UI look pretty.

> Our approach will focus on best practices in UI development so that your application looks good and works well on any Android device, regardless of screen size and resolution. We're going to develop a single app that will look great on phones, tablets, and TVs.

Networking and multithreading
> Once we have a working screen, we will want to post the user input to the cloud service. For that purpose, we are going to use a library to help us with the Twitter API web service calls.

Making the network calls is subject to the unpredictable nature of the network. To address that, we will introduce multithreading in Android and explain how to develop an app that works well regardless of external circumstances.

Debugging Android apps

A few things are going to go wrong in this section of the book. This is by design, because debugging is a normal part of application development. We'll show you how to use the Android SDK tools to quickly find and fix problems. Debugging will become second nature to you.

Fragments

Android 3.0 called for a newer approach to the user interface. The need to handle multiple screen sizes and orientations led to the introduction of fragments. We will tackle this UI framework by taking what we have done before and converting them over to this new approach.

Part 2: Intents, ActionBar, and More

This part, covered in Chapter 9, is about using Android intents as a way to connect multiple parts together. At the end of this part, your Yamba application will have two screens: one for status updates and the other for setting up the preferences. At this point, Yamba is configurable for various users and starts being a useful app. The elements we'll create at this stage are the activity, the menu system, and intents to glue them all together:

The Preference activity

First, we'll create the screen as an activity, one of Android's basic building blocks. You will see the steps involved and understand what it takes to create new screens.

Intents, ActionBar, and menu system

Next, we'll need a way to get to that screen. For that purpose, we'll introduce Action Bar as a menu system in Android and show how it works. You will also learn about intents and how to send these to open up a specific activity.

Filesystem

Finally, we'll learn about the filesystem on a typical Android device. You will gain a deeper understanding of how the operating system is put together, and you will also learn more about Android security.

Part 3: Android Services

This part, covered in Chapter 10, introduces background services. By the end of this part, your Yamba application will be able to periodically connect to the cloud and pull down your friends' status updates:

Services

Android services are very useful building blocks. They allow a process to run in the background without requiring any user interface. This is perfect for Yamba, because we'll have an update process connect to the cloud periodically and pull the data. In this section, you will also learn about multithreading considerations as they apply to background services.

Intent services

These are a convenient way to run a task off the main thread so that thread can continue to handle user interaction. We'll use one to get updates from Twitter.

Part 4: Content Providers

We now have the data from our refresh service, so we need a place to store it. In this part, covered in Chapter 11, we'll introduce you to Android's support for content providers, and data sources in general. By the end of that chapter, our data from the cloud will be persisted in the database:

SQLite and Android's support

Android ships with a built-in database called SQLite. In addition to this cool little database, the Android framework offers a rich API that makes SQLite easier for us to use. In this section, you will learn how to use SQLite and the API for it. You do not have to be a SQL buff to understand what is going on, but some basic understanding of SQL always helps.

The dbHelper class

To let you invoke the most common database operations without using SQL, Android rovides a class with the common `insert()`, `query()`, `update()`, and de `lete()` operations.

ContentProvider

We'll implement a new Android component to hold cached data and connect our app to it.

Part 5: Lists and Adapters

It might sound like we're back in UI mode, but lists and adapters are more organizational aids than user interface elements in Android. They form very powerful components that allow our tiny UI to connect to very large datasets in an efficient and scalable manner. In other words, users will be able to use Yamba in the real world without any performance hits in the long run.

Currently the data is all there in the database, but we have no way to view it. In this part, covered in Chapter 12, the Yamba application will get the much-needed `TimelineAc tivity` and a way for the user to see what his friends are chatting about online.

Part 6: Broadcast Receivers

Here we develop a third activity, doing so in multiple stages. First, we'll use our existing knowledge of the Android UI and put something together. It will work, but will not be as optimal as it could be. Finally, we'll get it right by introducing Lists and Adapters to the mix to use them to tie the data to our user interface.

In this part, covered in Chapter 13, we'll equip Yamba with receivers so it can react to events around it in an intelligent way. For that purpose, we'll use broadcast receivers. We show how to use Android permissions to make sure other people can't post statuses under the user's name:

Boot and network receivers
> In our example, we want to start our updates when the device is powered up. We also want to stop pulling the data from the cloud when the network is unavailable, and start it again only when we're back online. This goal will introduce us to one type of broadcast receiver.

Timeline receiver
> This type of receiver will exist only at certain times. Also, it won't receive messages from the Android system, but from other parts of our own Yamba application. This will demonstrate how we can use receivers to put together loosely coupled components in an elegant and flexible way.

Permissions
> At this point in the development process you know how to ask for system permissions, such as access to the Internet or filesystem. In this section, you'll learn how to define your own permissions and how to enforce them. After all, Yamba components might not want to respond to any other application for some Yamba-specific actions.

Part 7: App Widgets

In this part, covered in Chapter 14, we'll look at how to use Android app widgets to create a home screen widget that displays the latest tweets:

Android widgets
> But who will remember to pull up our app? To demonstrate the usefulness of our new status data, we'll put together an app widget. App widgets are those little components that the user can put on the home screen to see weather updates and such.

We'll create a widget that will pull the latest status update from the Yamba database via the status data content provider and display it on the home screen.

Part 8: Networking and the Web (HTTP)

Up till now we have provided the underlying communication piece to our example application via a library. Here we want to take a brief step back and talk about how this communication is done and what Android's APIs provide to communicate via HTTP.

Part 9: Live Wallpaper and Handlers

As a final piece to the application, we wanted to provide some more interaction at a system level. One of those ways is a fun, recently added concept in Android called Live Wallpaper, which runs on the home screen of the device. We build out a basic Live Wallpaper that interacts with the user and displays the messages communicated through the backend service. We also cover an important class called the Handler that enables another means to interact with the main UI thread from a different thread.

Summary

This chapter is intended as a road map for the next eight chapters. By the end of all these iterations, you will have built a medium-size Android app from scratch. Even more, you will understand various constructs and how to put them together into a meaningful whole. The hope is that you'll start developing a way of *thinking* in Android.

## CHAPTER 7

Android User Interface

In this chapter, you will learn how to build a user interface in Android. You will create your first activity, create an XML layout for it, and see how to connect it to your Java code. You will learn about views (a.k.a. widgets) and layouts, and learn how to handle Java events, such as button clicks. Additionally, you'll add support for a Twitter-like API into your project as an external *.jar* file so your app can make web service calls to the cloud.

By the end of this chapter, you will have written your own Twitter-like Android app. The app will feature a single screen that will prompt the user for her current status update and post that update online.

Two Ways to Create a User Interface

There are two ways to create a user interface (UI) in Android: declaratively and programmatically. They are quite different but often are used together to get the job done.

Declarative User Interface

The declarative approach involves using XML to declare what the UI will look like, similar to creating a web page using HTML. You write tags and specify elements to appear on your screen. If you have ever handcoded an HTML page, you did pretty much the same work as creating an Android screen.

One advantage of the declarative approach is that you can use WYSIWYG tools. Some of these tools ship with the Eclipse Android Development Tools (ADT) extension, and others come from third parties. Additionally, XML is fairly human-readable, and even people who are unfamiliar with the Android platform and framework can readily determine the intent of the user interface.

The disadvantage of a declarative UI approach is that you can get only so far with XML. XML is great for declaring the look and feel of your user interface, but it doesn't provide a good way of handling user input. That's where the programmatic approach comes in.

Programmatic User Interface

A programmatic user interface involves writing Java code to develop the UI. If you have ever done any Java AWT or Java Swing development, Android is pretty much the same in that respect. It is similar to many UI toolkits in other languages as well.

Basically, if you want to create a button programmatically, you have to declare the button variable, create an instance of it, add it to a container, and set any button properties that may make sense, such as color, text, text size, background, and so on. You probably also want to declare what the button does once it's clicked, so that's another piece of code. All in all, you end up writing quite a few lines of Java.

Everything you can do declaratively, you can also do programmatically. But Java also allows you to specify what happens when that button is actually clicked. This is the main advantage of a programmatic approach to the user interface.

The Best of Both Worlds

So which approach to use? The best practice is to use both. Use the declarative (XML) approach to declare everything about the user interface that is static, such as the layout of the screen, all the widgets, etc. Then switch to a programmatic (Java) approach to define what goes on when the user interacts with the various widgets in the user interface. In other words, you'd use XML to declare what the "button" looks like and Java to specify what it does.

 Note that there are two approaches to developing the actual user interface, but at the end of the day, all the XML is actually "inflated" into Java memory space as if you actually wrote Java code. So it's only Java code that runs.

Views and Layouts

Android organizes its UI elements into *views* and *layouts*. Everything you see, such as a button, label, or text box, is a view. Layouts organize views, such as grouping together a button and label or a group of these elements.

If you have prior experience with Java AWT or Swing, layouts are similar to Java containers and views are similar to Java components. Views in Android are sometimes referred to as *widgets*.

Don't confuse widgets in the Android UI with App Widgets. The latter are miniature application views that can be embedded in other applications (such as the home screen application). Here, we are referring to widgets as the views in our activities.

A layout can contain other children. Those children can furthermore be layouts themselves, allowing for a complex user interface structure (Figure 7-1). A layout is responsible for allocating space for each child.

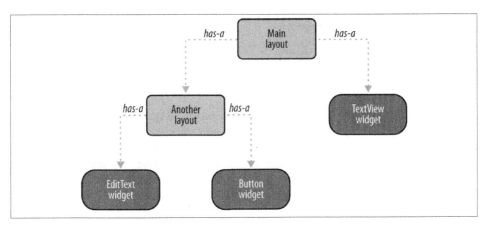

Figure 7-1. Layouts and views relationship

Some of the most common layouts follow. There are others, but they are used less frequently.

LinearLayout

LinearLayout is one of the simplest and most common layouts (see Figure 7-2). It simply lays out its children next to one another, either horizontally or vertically. The order of the children matters. As LinearLayout asks its children how much space they need, it allocates the desired space to each child in the order it is added. So if an "older" child comes along and asks for all the space on the screen, there won't be much left for the subsequent widgets in this layout.

One important property for LinearLayout is layout_orientation. Its valid options are vertical or horizontal.

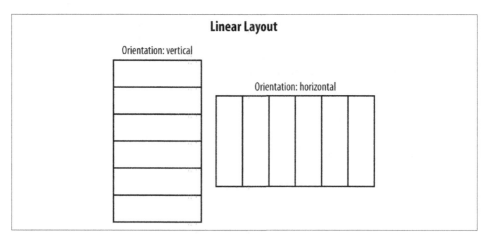

Figure 7-2. LinearLayout

 Although LinearLayout is probably the simplest and most commonly used layout, it is not always the best choice. A good rule of thumb is that if you start to nest multiple LinearLayouts, you should probably use a different layout, such as RelativeLayout. Too many nested layouts can have major consequences on the time needed to inflate the UI and on overall CPU and battery consumption.

TableLayout

TableLayout lays out its children in a table, and the views it contains are TableRow widgets (see Figure 7-3). Each TableRow represents a row in a table and can contain other UI widgets. TableRow widgets are laid out next to each other horizontally, like LinearLayout with a horizontal orientation.

For those familiar with HTML, TableLayout is similar to the <table> element, and TableRow is similar to the <tr> element. Whereas HTML also offers <td> to represent each cell in the table, Android determines the columns dynamically based on the number of views you add to a TableRow.

An important property for TableLayout is stretch_columns, indicating which column of the table to stretch. You can also use * to stretch all columns.

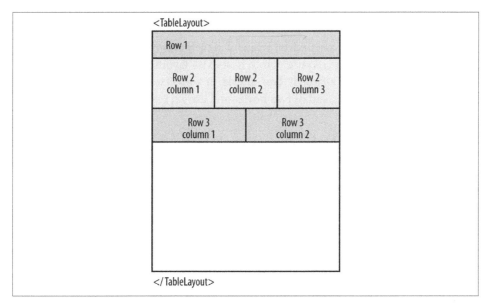

Figure 7-3. TableLayout

FrameLayout

FrameLayout places its children on top of each other so that the latest child is covering the previous one, like a deck of cards (see Figure 7-4). This layout policy is useful for tabs, as one example. FrameLayout is also used as a placeholder for other widgets that will be added programmatically at some later point in time.

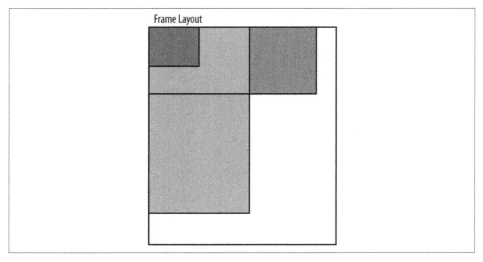

Figure 7-4. FrameLayout

RelativeLayout

RelativeLayout lays out its children relative to each other (see Figure 7-5). It is very powerful because it doesn't require you to nest extra layouts to achieve a certain look. For instance, if you have a two-by-two matrix of widgets, you can lay them out in a single RelativeLayout instead of two horizontal LinearLayouts within a vertical LinearLayout.

By streamlining the number of layouts you use, RelativeLayout can minimize the total number of widgets that need to be drawn, thus improving the overall performance of your application. On the other hand, RelativeLayout adds a bit of complexity by requiring each child view to have an ID so that you can position it relative to other children.

Once hard to use, RelativeLayout is becoming the most versatile and efficient layout of them all. Android Tools for Eclipse, as well as Android Studio, both have very good support for visually setting the relationship constraints of the view's layout using RelativeLayout.

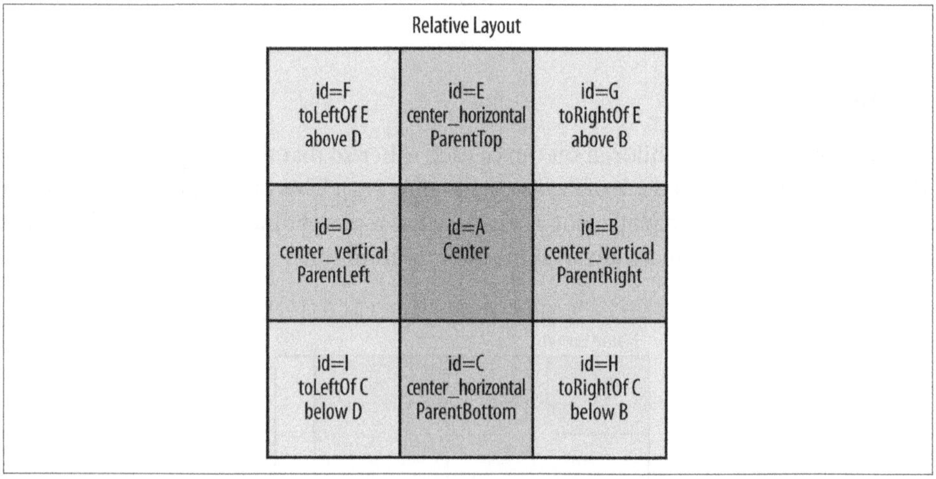

Relative Layout		
id=F toLeftOf E above D	id=E center_horizontal ParentTop	id=G toRightOf E above B
id=D center_vertical ParentLeft	id=A Center	id=B center_vertical ParentRight
id=I toLeftOf C below D	id=C center_horizontal ParentBottom	id=H toRightOf C below B

Figure 7-5. RelativeLayout

Starting the Yamba Project

We are about to start our Yamba project. So fire up Eclipse and click File → New → Android Application Project.

You will get a dialog window asking you about your new Android project (see Figure 7-6). Let's explain again all the significant fields:

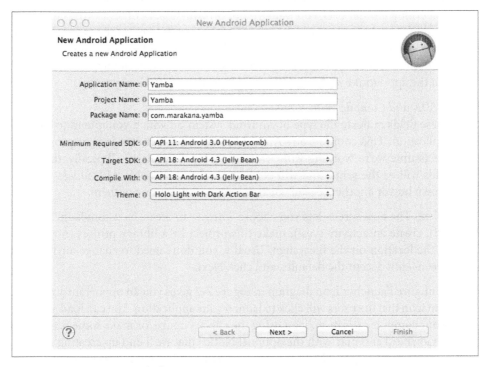

Figure 7-6. New Project dialog

Application Name
> This is the name of your application as generally visible to the users. It can be pretty much any text, and you can easily change it later. We'll simply call our app "Yamba."

Project Name
> The name under which Eclipse organizes our project. It is a good idea not to use any spaces in your project name. This makes it easier to access from the command line later. Enter **Yamba** here.

Package Name

This field designates a Java package, so it needs to adhere to Java package naming conventions (*http://en.wikipedia.org/wiki/Java_package#Package_naming_conven tions*). In a nutshell, you want to use the reverse of your domain name for your package. We're going to use "com.marakana.android.yamba" here.

Min Required SDK Version

Represents the minimum version of Android SDK that must be installed on the device for it to run this particular application. At this point, this is a business choice concerning how far back you want to support your app. In other words, should this app be able to run on very old devices? There's a certain programming cost involved in backward support. An analogy to this would be building a website that has to work on IE6, where you have to deal with a lot of quirks. In our case, we choose that the app should work on API level 11 (Android 3.0, Honeycomb).

Target SDK and Compile With

These fields indicate the type of Android system on which you intend to run this application. This could be any Android platform, either standard or proprietary. We assume we're working with Android 4.3 (API level 18). Typically, these two fields will be the same, and will point to the latest available API, unless there's a known bug or another reason you'd downgrade either one of them.

Click Next. The next screen lets you choose whether to create a custom launcher icon (we do!), create an activity (yes!), make this project be a library project (nope), and choose the location on the filesystem. Usually, you don't need to change anything on this screen—just accept the defaults, and click Next.

The Configure Launcher Icon diagram in Figure 7-7 gives you an opportunity to create a custom icon that the users will click to launch your application. You can load an image from your computer, create an icon using the clip art image, or make some text be that image. Go ahead and play with the options! Notice that we'll end up creating not one, but four different images. They are all the same image, but with different sizes. This is because Android runs on thousands of different devices with a variety of screen densities. Some older devices have about 160 dots per inch (dpi), whereas the newer ones are pushing 400 dpi. Depending on the type of the display, Android OS will pick the right image to render.

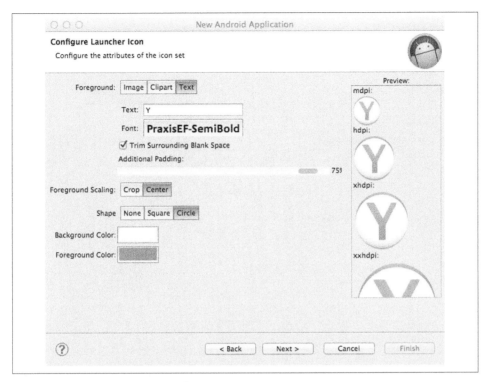

Figure 7-7. Configure the Launcher Icon

Once you are happy with that image, click Next. On the Create Activity screen, choose to create an activity, and leave it at just the Blank Activity. We'll start with an empty canvas and build on top of that. Click Next.

On the Blank Activity screen in Figure 7-8, pick the name of the activity. While you're here, keep in mind that your activity will be represented by a Java class. Thus, for the activity name, you must adhere to Java class naming conventions (*http://en.wikipe dia.org/wiki/Naming_convention_(programming)#Java*). Doing that simply means using upper camel case (*http://en.wikipedia.org/wiki/CamelCase*). We'll enter **StatusActivity** here.

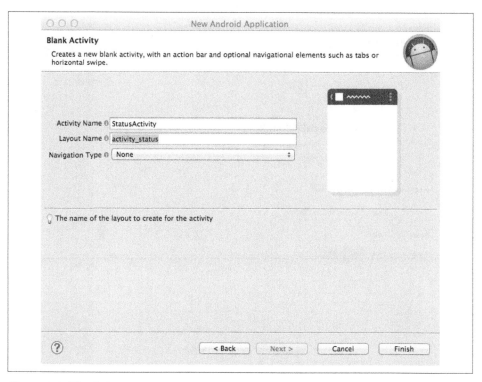

Figure 7-8. Blank Activity setup

Typically, an activity will have a layout file associated with it. This file is an XML file. While we have a strong naming convention in Java, the XML resources could be named both using the camel case as well as the underscores. In this case, go with what Eclipse ADT is suggesting—"activity_status"—so go ahead and accept that.

Click Finish. Your Yamba project should now appear in Eclipse's Package Explorer, as shown in Figure 7-9.

Figure 7-9. Eclipse with the boilerplate Yamba project

The StatusActivity Layout

Let's start by designing the user interface for our screen, where we'll enter the new status and click a button to update it.

To start creating this screen, open the *res/layout/activity_status.xml* file. Eclipse will recognize this file and open it using a graphical layout. As a matter of fact, this file should have already been opened when you created the project initially.

Notice that you can toggle between the graphical layout and the actual generated code via the tabs at the bottom of the window. Think of this tool as a Dreamweaver or similar HTML authoring tool where you can easily create a UI, and the tool generates the underlying code.

Notice that on the left side of the tool you have the Palette with various widgets. We're going to use these widgets in the graphical mode to create our UI. But first, go ahead and select the "Hello world!" text in the window, and delete it. Now you have a blank screen to work with.

Our StatusActivity screen will have these components:

- Button to post a tweet
- Text area to type the text of the message
- Box that contains it all, also known as a layout

The layout is already given to us. This is the big white area. It's of type `RelativeLayout`. We discussed the layout types earlier in this chapter.

We'll begin with a button to click to update the status. This will be a `Button` widget. Locate it under the Form Widgets section of the Palette and drag it to the top-right corner of your screen. The button will sort of glue to that corner.

Next, we need a big text area to type our 140-character status update. We'll use an `EditText` widget for this purpose. You can locate this widget under the Text Fields section of the Palletes. Although there are a few choices, most of them are the same element with just a different visual appearance. Pick Multiline Text for our case. Drag it just below the button, "gluing" it to the left margin.

To make the text area large, right-click it and choose Layout Width→Match Parent. Do the same for the height.

Next, let's update the text. Right-click the button and choose "Edit Text." Now, you could simply change the word "Button" to something else, but instead, choose "New String" (Figure 7-10). Under "New R.string." type **button_tweet**. This will be the ID of this particular piece of text. Under String type the actual English value, for example the word "Tweet," then click OK, and OK again in the underlying window.

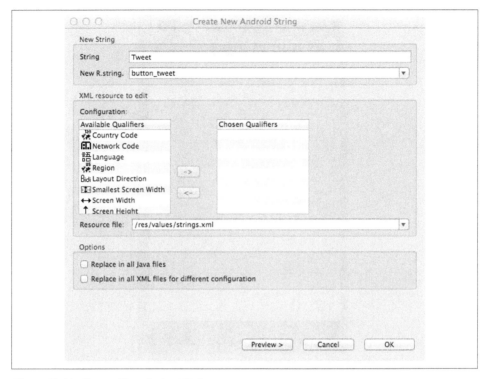

Figure 7-10. Create New String dialog

Repeat the same steps for the big text area, but this time around right-click and choose "Edit Hint." For the ID of the new string, let's pick "hint_status."

At this point, our screen looks like Figure 7-11.

Figure 7-11. StatusActivity graphical layout

There is one more thing we should do: initialize the IDs of elements we'll care about programmatically. In our case, we will want to be able to lookup the button and the text area in Java. To do that, we want to assign them meaningful IDs. Eclipse already created IDs such as *button1* and *editText1*, but those don't tell the story. To change them, right-click the button, pick "Edit ID," then type **buttonTweet** and click OK. Do the same for the status text area and type **editStatus** for the ID.

You could also peek at the actual generated code by clicking the tab at the bottom of the screen. Example 7-1 contains the source code for our StatusActivity layout.

Example 7-1. The res/layout/activity_status.xml file

```
<RelativeLayout xmlns:android="http://schemas.android.com/apk/res/android"
    xmlns:tools="http://schemas.android.com/tools"
    android:layout_width="match_parent"
    android:layout_height="match_parent"
    android:paddingBottom="@dimen/activity_vertical_margin"
    android:paddingLeft="@dimen/activity_horizontal_margin"
    android:paddingRight="@dimen/activity_horizontal_margin"
    android:paddingTop="@dimen/activity_vertical_margin"
    tools:context=".StatusActivity" >
```

```
<Button
    android:id="@+id/buttonTweet"
    android:layout_width="wrap_content"
    android:layout_height="wrap_content"
    android:layout_alignParentRight="true"
    android:layout_alignParentTop="true"
    android:layout_marginRight="17dp"
    android:text="@string/button_tweet" />

<EditText
    android:id="@+id/editStatus"
    android:layout_width="match_parent"
    android:layout_height="match_parent"
    android:layout_alignParentLeft="true"
    android:layout_below="@+id/buttonTweet"
    android:layout_marginTop="10dp"
    android:ems="10"
    android:hint="@string/hint_status"
    android:inputType="textMultiLine" >

    <requestFocus />
</EditText>

</RelativeLayout>
```

 Though you could just copy-paste this file into your project, note that it references couple of string resources in another file. So just copying this file may not work. More on strings later in this chapter.

Important Widget Properties

The properties you are most likely to use regularly are:

layout_height *and* layout_width

Define how much space this widget is asking from its parent layout to display itself. Although you could enter a value in pixels, inches, or something similar, that is not a good practice. Because your application could run on many different devices with various screen sizes, you want to use relative size for your components, not an absolute size. So the best practice is to use either match_parent or wrap_content for the value. match_parent means that your widget wants all the available space from its parent. wrap_content means that it requires only as much space as it needs to display its own content. Prior to API 8, match_parent was known as fill_par ent. Some developers still use the older name, which still works.

layout_weight
> The layout weight is a number between 0 and 1 that implies the weight of our layout requirements. For example, if our `Status EditText` had a default layout weight of 0 and required a layout height of `fill_parent`, the Update button would be pushed out of the screen because Status and its request for space came before the button. However, when we set the Status widget's layout weight to 1, we are saying we want all available space along the dimension of height, but will yield to any other widget that also may need space, such as the Update button. Note that `layout_weight` doesn't apply to a relative layout.

layout_gravity
> Specifies how this particular widget is positioned within its parent layout, both horizontally and vertically. Values can be `top`, `center`, `left`, and so on.

gravity
> Specifies how the content of this widget is positioned within the widget. The difference between `gravity` and `layout_gravity` is explained after this list.

text
> Not all widgets have this property. Some widgets with text include `Button`, `Edit Text`, and `TextView`. This property simply specifies the text to show in the widget. However, it is not a good practice to just enter the text, because then your layout will work in only one locale/language. Best practice is to define all text in a *strings.xml* resource file and refer to a particular string using this notation: `@string/titleStatusUpdate`.

id
> `id` is simply the unique identifier for this particular widget in a particular layout resource file. Not every widget needs an `id`, and we recommend removing unnecessary `ids` to minimize clutter. But widgets that you need to manipulate later from Java do need `ids`. An `id` has the format `@+id/`*someName,* where *someName* is whatever you want to call your widget. The naming convention we use is to put in the type followed by the name, for example, `@+id/buttonUpdateStatus`.

The difference between `gravity` and `layout_gravity` is subtle, but will be easy to understand after you try them out. `gravity` controls what happens inside a widget: you would use it, for example, to left-justify or center text within the widget. `layout_grav ity` deals with the relationship between the widget and its parent. For instance, if our `Title TextView` had its width set to `wrap_content`, we could specify `layout_gravi ty="center_horizontal"` to center it horizontally within its parent LinearLayout. If you've worked with web pages, you can consider `gravity` to be like padding in CSS, whereas `layout_gravity` is like margins.

Strings Resource

Android tries hard to keep data in separate files. So layouts are defined in their own resources, and all text values (such as button text, title text, etc.) should be defined in their own file called *strings.xml*. This allows you later to provide multiple versions of string resources for various languages, such as English, Japanese, or Russian.

As usual, Eclipse ADT provides an easy editor to manage known resource types, which include strings. The editor window looks like Figure 7-12.

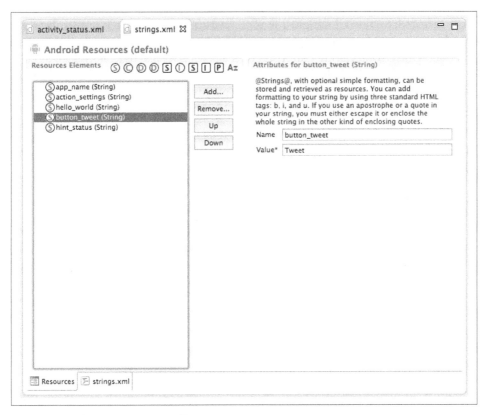

Figure 7-12. Strings resources editor

To see the underlying source code, choose the "strings.xml" tab at the bottom of that editor window. Example 7-2 shows what our *strings.xml* file looks like at this point.

Example 7-2. The res/values/strings.xml file

```
<?xml version="1.0" encoding="utf-8"?>
<resources>

    <string name="app_name">Yamba</string>
```

```
<string name="action_settings">Settings</string>
<string name="hello_world">Hello world!</string>
<string name="button_tweet">Tweet</string>
<string name="hint_status">What\'s going on?</string>
```

```
</resources>
```

The file simply contains sets of name-value pairs.

The StatusActivity Java Class

Now that we have our UI designed in XML, we are ready to switch over to Java. Remember from earlier in this chapter that Android provides two ways for building a user interface. One is by declaring it in XML, which is what we just did, and we got as far as we could (for now). The other one is to build it programmatically in Java. We also said earlier that the best practice is to get as far as possible in XML and then switch over to Java.

Our Java class for this is `StatusActivity.java`, and the Eclipse New Project dialog has already created the stub for this class. The class is part of the `com.marakana.an droid.yamba` Java package, and as such is part of that directory.

Inflating XML to Java

As with all main building blocks in Android, such as activities, services, broadcast receivers, and content providers, you usually start by subclassing a base class provided by the Android framework and overriding certain inherited methods. In this case, we subclass Android's `Activity` class and override its `onCreate()` method. As you recall, activities have a certain life cycle or state machine (see "Activity Life Cycle" on page 64) through which they go. We as developers do not control what state the activity is in, but we do get to say what happens during a transition to a particular state. In this case, the transition we want to override is the `onCreate()` method that the system's Activity Manager invokes when the activity is first created (i.e., when it goes from a starting to a running state). This sort of programming, when we subclass a system class and fill in the blanks, is also known as the Template pattern (*http://en.wikipedia.org/wiki/ Template_method_pattern*).

In addition to doing some standard housekeeping, our `onCreate()` will carry out two major tasks that the application needs done just once, at the beginning: set up our button so it responds to clicks, and connect to the cloud.

Notice that `onCreate()` takes a `Bundle` as a parameter. This is a small amount of data that can be passed into the activity once it is being shut down so that the new instance of this activity can recreate its original state. This is a common case when rotating the screen, in which case the activity typically gets reinitialized. The data provided in a

`Bundle` is typically limited to basic data types; more complex ones need to be specially encoded. For the most part, we're not going to be using `Bundle` in our Yamba example, because the application has no real need for it.

Keep in mind that whenever you override a method, you first want to make a call to the original method provided by the parent. That's why we have a `super.onCreate()` call here.

So once you subclass the framework's class, override the appropriate method, and call `super`'s method in it, you are still back where you started: your code does the same thing the original class did. But now you have a placeholder where you can add your own code.

The very first thing you typically do in an activity's `onCreate()` is to load the UI from the XML file and *inflate* it into the Java memory space. In other words, write some Java code that opens up your XML layout file, parses it, and for each element in XML, creates a corresponding Java object in your memory space. For each attribute of a particular XML element, this code will set that attribute on your Java object. The line of code that does all this is `setContentView(R.layout.activity_status);`.

 Remember that the `R` class is the automatically generated set of pointers that helps connect the world of Java to our world of XML and other resources in the */res* folder. Similarly, `R.layout.activity_sta tus` points to our */res/layout/activity_status.xml* file.

This `setContentView()` method does a lot of work, in other words. It reads the XML file, parses it, creates all the appropriate Java objects to correspond to XML elements, sets object properties to correspond to XML attributes, sets up parent/child relationships between objects, and overall inflates the entire view. At the end of this one line, our screen is ready for drawing.

The Eclipse boilerplate code also includes the `onCreateOptionsMenu()` method. For now, we'll ignore this method. We'll get back to it in "The Action Bar" on page 148. Example 7-3 shows the code that ADT produces once we run through this New Project wizard.

Example 7-3. StatusActivity.java, boilerplate code given by ADT

```
package com.marakana.android.yamba;

import android.os.Bundle;
import android.app.Activity;
import android.view.Menu;

public class StatusActivity extends Activity { // ❶
```

```
@Override
protected void onCreate(Bundle savedInstanceState) { // ❷
    super.onCreate(savedInstanceState); // ❸
    setContentView(R.layout.activity_status); // ❹
}


@Override
public boolean onCreateOptionsMenu(Menu menu) {
// Inflate the menu; this adds items to the action bar if it is present.
    getMenuInflater().inflate(R.menu.status, menu);
    return true;
}

}
```

❶ Every activity is a subclass of the Activity class.

❷ We override the onCreate() method in order to add some specific logic.

❸ Remember to call the super in all the life cycle methods.

❹ This is the main work of this entire code. This is where the activity loads up the XML from *res/layout/activity_status.xml*. For each element in that XML file, the call creates the corresponding Java object of the same class as the name.

Initializing Objects

Once we inflate the objects into the Java memory space, we have to *find* the objects that we actually care about and assign them to Java variables. To do that, we declare these variables, usually as private and class-global. Next, we use findViewById() method to look them up:

```
...
    private EditText editStatus; // ❶
    private Button buttonTweet;

    @Override
    protected void onCreate(Bundle savedInstanceState) {
        super.onCreate(savedInstanceState);
        setContentView(R.layout.activity_status);

        editStatus = (EditText) findViewById(R.id.editStatus); // ❷
        buttonTweet = (Button) findViewById(R.id.buttonTweet);
    }
...
```

❶ This is where we define the Java variable as private and class-global.

❷ We look up the actual object from the Java memory space using findViewById(), but only after inflating the XML using setContentView().

Handling User Events

Your objects are not the only ones that define methods and respond to external stimuli. Android's user interface objects do that, too. Thus, you can tell your `Button` to execute certain code when it's clicked. To do that, you need to define a method named on Click() and put the code there that you want executed. You also have to run the setOnClickListener method on the `Button`. Pass this as an argument to setOnClick Listener, because your current object (the activity) is where you define onClick(). Example 7-4 shows our first version of *StatusActivity.java*, with some additional explanation following the code.

Example 7-4. StatusActivity.java, version 1

```
package com.marakana.android.yamba;

import android.app.Activity;
import android.os.Bundle;
import android.util.Log;
import android.view.Menu;
import android.view.View;
import android.view.View.OnClickListener;
import android.widget.Button;
import android.widget.EditText;

public class StatusActivity extends Activity implements
                                OnClickListener { //❶
    private static final String TAG = "StatusActivity";
        private EditText editStatus;
        private Button buttonTweet;

        @Override
        protected void onCreate(Bundle savedInstanceState) {
                super.onCreate(savedInstanceState);
                setContentView(R.layout.activity_status);

                editStatus = (EditText) findViewById(R.id.editStatus);
                buttonTweet = (Button) findViewById(R.id.buttonTweet);

                buttonTweet.setOnClickListener(this); // ❷
        }

        @Override
        public void onClick(View view) { // ❸
                String status = editStatus.getText().toString(); // ❹
                Log.d(TAG, "onClicked with status: " + status); // ❺
        }

        @Override
        public boolean onCreateOptionsMenu(Menu menu) {
// Inflate the menu; this adds items to the action bar if it is present.
                getMenuInflater().inflate(R.menu.status, menu);
```

```
            return true;
        }

    }
```

❶   To make `StatusActivity` capable of being a button listener, it needs to implement the `OnClickListener` interface.

❷   Register the button to notify `this` (i.e., `StatusActivity`) when it gets clicked.

❸   The method that is called when button is clicked, as part of the `OnClickListen er` interface.

❹   We look up the value of the actual text of the status in the UI.

❺   Use the log system to print out the value so we know this is working.

Logging Messages in Android

Android offers a system-wide logging capability. You can log from anywhere in your code by calling `Log.d(TAG, message)`, where *TAG* and *message* are some strings. `TAG` should be a tag that is meaningful to you given your code. Typically, a tag would be the name of your app, your class, or some module. Good practice is to define `TAG` as a Java constant for your entire class, such as:

```
private static final String TAG = "StatusActivity";
```

 When you add logging to your app, before your code will compile, you need to import the Log class. Eclipse has a useful feature under Source → Organize Imports, or Ctrl-Shift-O for short. Usually, this feature will automatically organize your import statements. However, in the case of Log, often there is a conflict because there are multiple classes named Log. This is where you have to use your common sense and figure it out. In this case, the ambiguity is between the Android Log and Apache Log classes, so the choice should be easy.

Note that `Log` takes different severity levels. `.d()` is for debug level, but you can also specify `.e()` for error, `.w()` for warning, or `.i()` for info. There's also a `.wtf()` severity level for errors that should never happen. (It stands for "what a terrible failure," in case you were wondering.) Eclipse color codes log messages based on their severity level.

LogCat

The Android system log is outputted to LogCat, a standardized system-wide logging mechanism. LogCat is readily available to all Java code. The developer can easily view the logs and filter their output based on severity, such as debug, info, warning, or error, or based on custom-defined tags. As with most things in Android development, there are two ways to view the LogCat: via Eclipse or via the command line.

To view LogCat in Eclipse, you need to open the LogCat View (see Figure 7-13). Typically, Eclipse will automatically open this view for you upon running of the application, but you can also manually open it up by going to Window → Show View → Other → Android → LogCat.

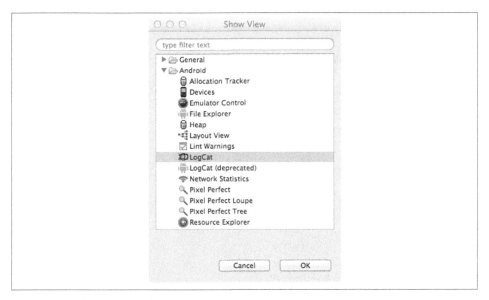

Figure 7-13. LogCat in Eclipse

You can define filters for LogCat as well. Click the little green plus button, and the LogCat Filter dialog will come up. You can define a filter based on a tag, severity level, or process ID. This will create another window within LogCat that shows you only the log entries that match your filter.

Compiling Code and Building Your Projects: Saving Files

After you make changes to your Java or XML files, make sure you save them before moving on. Eclipse builds your project automatically every time you choose File → Save or press Ctrl-S. So it is important to save files and make sure you do not move to another file until the current file is fine. You will know your file is fine when there are no little red *x* symbols in your code and the project builds successfully. Because Java depends on XML and vice versa, moving to another file while the current one is broken just makes it even more difficult to find errors.

Java errors typically are easy to find because the little red *x* in the code navigates you straight down to the line number where the error occurred. By putting your mouse right on that error, Eclipse will tell you what the error is and will also offer you some possible fixes. This Eclipse feature is very useful and is analogous to the spellchecker in a word processor.

At this point, you can run your application. Right-click the Yamba project in the Package Explorer, and choose Run As → Android Application (Figure 7-14).

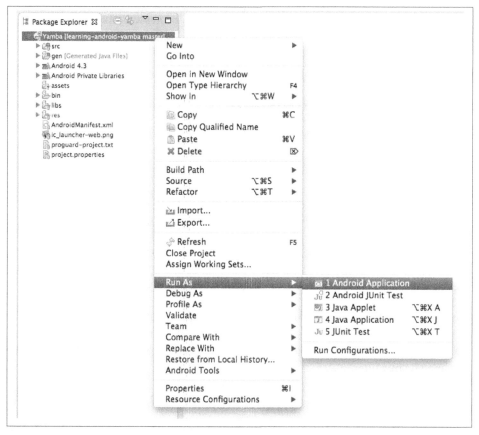

Figure 7-14. Running the app

At this point, your application should show up on your emulator (Figure 7-15), or your real device for that matter.

Figure 7-15. Yamba on the emulator

Next, we're going to start working on getting our application to post to the cloud via a web service call.

A Brief History of the Twitter API and This Book's Substitute

In 2006, Twitter introduced its first API in response to those who were simply scraping the site in order to get the tweets. This API is based on the RESTful interface serving JSON and XML data.

Twitter also used to support simple authentication, where all you'd need to provide to authenticate your request with the web service was your username and password. Twitter has since moved to *OAuth* authentication (*http://oauth.net/*)—a smarter way to give apps access to your account without giving them your login credentials.

In the early days, Yamba applications did indeed work with the real Twitter.com service. Since then, Twitter has changed its original API, making it harder for third-party apps to use the service. Specifically, the new API replaced simple authentication with the more secure but also more complex OAuth authentication. It has also introduced *rate limiting*, which makes it harder for multiple people to use one account at the same IP, thus creating friction for classroom learning environments.

OAuth works based on an exchange of tokens. Though in general this is a better way to authenticate, we felt that for learning purposes it threw the flow of this tutorial-style approach off on a tangent that is not all that relevant at this point in learning Android. There are numerous articles written about how to use OAuth with your Android app, including these:

- OAuth in Android (*http://marakana.com/forums/android/examples/312.html*)
- Authentication with OAuth 2.0 (*http://bit.ly/1gc7WoJ*)

Instead of depending on the ever-changing world of the Twitter API, we have created a Twitter-like service at *http://yamba.marakana.com*. This service, built using the Status.net Twitter-compatible API (*http://status.net/wiki/Twitter-compatible_API*), implements the Twitter API 1.0 and captures the spirit of what we intend to showcase in our Yamba example without all the complications of the real Twitter service.

Adding the Twitter API Library

To process web service API calls, you need an HTTP library and an XML parser. Although a number of open source libraries provide general HTTP and XML parsing capabilities, including a couple of Android-specific options, you need to know more about Java networking than Android programming to implement web service calls.

To make our life with web services and the Twitter API easier, we're going to use a third-party library, YambaClientLib (*https://github.com/marakana/YambaClientLib*), which we created. This library contains a simple Java class that interacts with the online service and abstracts all the intricacies of making network calls and passing the data back and forth. Again, if there were no higher-level library for what we needed to do, we could have used standard Java networking libraries to get the job done. It just would have been more work, and that work is not directly relevant to learning Android.

The YambaClientLib library has been designed specifically for the purposes of this book. It has been stripped down to the bare essentials, making it easy for you to peek at its source code and see the its inner workings, if you care to do so. It also supports Twitter's older API that allows for simple authentication (username and password) versus the new OAuth authentication.

After you download this library, you can put it inside your project in Eclipse. Simply drag the *yambaclientlib.jar* file and drop it in the *libs* folder of your Eclipse project in the Package Manager window. This makes the file part of the project. More so, any *.jar* file dropped in this special *libs* folder of your project automatically becomes part of the project's `classpath`, where the Java compiler will look for the libraries it needs to load.

Updating the Manifest File for Internet Permission

Before this application can work, we must ask the user to grant us the right to use the Internet. Android manages security by specifying the permissions needed for certain dangerous operations. The user then must explicitly grant those permissions to each application when he first installs it. The user has the binary choice of granting all or no permissions requested by the application; there's no middle ground. Also, the user is not prompted about permissions when upgrading an existing app.

 Because we are running this application in debug mode and installing it via a USB cable, Android doesn't prompt us for permissions like it would the end user. However, we still must specify that the application requires certain permissions.

In this case, we want to ask the user to grant this application the INTERNET permission. We need Internet access to connect to the online service. So open up the *AndroidManifest.xml* file by double-clicking it. Note that Eclipse typically opens this file in a WYSIWYG editor with many tabs on the bottom. As always, you can make most of the changes to this file via this interface, but because Eclipse tools are limited and sometimes buggy, we prefer to go straight into the XML view of this file. So, choose the rightmost tab at the bottom that says_AndroidManifest.xml_, and add a <uses-permission> element within the <manifest> block (see Example 7-5).

Example 7-5. AndroidManifest.xml

```
<?xml version="1.0" encoding="utf-8"?>
<manifest xmlns:android="http://schemas.android.com/apk/res/android"
    package="com.marakana.android.yamba"
    android:versionCode="1"
    android:versionName="1.0" >

    <uses-sdk
        android:minSdkVersion="11"
        android:targetSdkVersion="18" />

    <!-- ❶ -->
    <uses-permission android:name="android.permission.INTERNET" />

    <application
        android:allowBackup="true"
        android:icon="@drawable/ic_launcher"
        android:label="@string/app_name"
        android:theme="@style/AppTheme" >
        <activity
            android:name="com.marakana.android.yamba.StatusActivity"
            android:label="@string/app_name" >
            <intent-filter>
```

```
            <action android:name="android.intent.action.MAIN" />

            <category android:name="android.intent.category.LAUNCHER" />
        </intent-filter>
      </activity>
   </application>

</manifest>
```

❶   Defines the `<uses-permission>` element for the `INTERNET` permission.

Threading in Android

A thread is a sequence of instructions executed in order. Although each CPU core can process only one instruction at a time, most operating systems are capable of handling multiple threads on multiple CPU cores, or interleaving them on a single CPU. Different threads need different priorities, so the operating system determines how much time to give each one if they have to share a CPU.

The Android operating system is based on Linux, and as such, is fully capable of running multiple threads at the same time. However, you need to be aware of how applications use threads in order to design your application properly.

Single Thread

By default, an Android application runs on a single thread. Single-threaded applications run all commands serially, meaning that no command starts until the previous one is done. Another way of saying this is that each call is *blocking*.

This single thread is also known as the *UI thread* because it's the thread that processes all the user interface commands. The UI thread is responsible for drawing all the elements on the screen as well as processing all the user events, such as touches on the screen, clicks of the button, and so on. Figure 7-16 shows the execution of our code on a single UI thread.

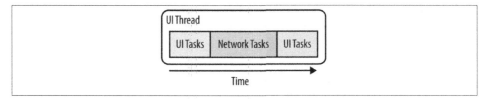

Figure 7-16. Single-threaded execution

The problem with running `StatusActivity` on the single thread comes with our network call to update the status. As with all network calls, the time it takes to execute is outside of our control. Our call to post to the cloud service is subject to all the network availability and latency issues. We don't know whether the user is on a super fast WiFi connection or is using a much slower protocol to connect to the cloud. In other words, our application cannot respond until the network call is completed.

Prior to Honeycomb (API 11), doing network calls on the UI thread was doable by default, but not recommended. Newer versions of Android OS will throw an exception, in this case, `android.os.NetworkOnMainThreadException`.

The Android system will offer to kill any application that is not responding within a certain time period, typically around five seconds for activities. This is known as the Application Not Responding dialog, or ANR for short (see Figure 7-17).

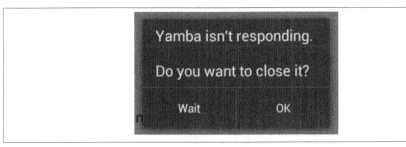

Figure 7-17. Application Not Responding dialog

Multithreaded Execution

A much better solution is to run the potentially long operations on a separate thread. When multiple tasks run on multiple threads at the same time, the operating system slices the available CPU time so that no one task dominates the execution. As a result, it appears that multiple tasks are running in parallel at the same time.

In our example, we could put the actual network call for updating our status in the cloud in a separate thread. That way our main UI thread will not block while we're waiting for the network, and the application will appear much more responsive. We tend to talk of the main thread as running in the *foreground* and the additional threads as running in the *background*. They're really all equal in status, alternating their execution on the device's CPU, but from the point of view of the user, the main thread is in the foreground because it deals with the UI. Figure 7-18 shows the execution of our code's two threads —the main UI thread, as well as the auxiliary thread we use to perform potentially long-running network calls.

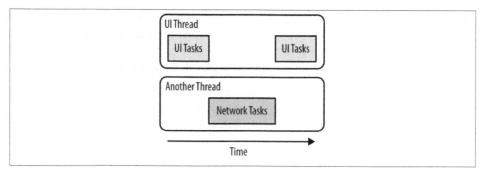

Figure 7-18. Multithreaded execution

There are several ways to accomplish multithreading. Java has a Thread class that allows for many of these operations. We could certainly use any of the regular Java features to put the network call in the background.

However, using the standard Java Thread class is somewhat problematic. Imagine that after we post to the cloud we want to notify the user of the success or failure of that operation. We would have to update the UI. But, in Android, a thread that didn't create the UI widget is not allowed to update the UI—this would not be thread-safe. We would need to synchronize these threads somehow, and that would be a job on its own.

Because this is a common task in Android, the framework provides the utility class AsyncTask specifically designed for this purpose.

AsyncTask

AsyncTask is an Android mechanism created to help handle long operations that need to report to the UI thread. To take advantage of this class, we need to create a new subclass of AsyncTask and implement its doInBackground(), onProgressUpdate(), and onPostExecute() methods. In other words, we are going to fill in the blanks for what to do in the background, what to do when there's some progress, and what to do when the task completes.

We'll extend our earlier example with an asynchronous posting to the cloud. The first part of Example 7-6 is very similar to the code in Example 7-4, but hands off the posting to the asynchronous task. A new AsyncTask does the posting in the background.

Example 7-6. StatusActivity.java, version 2

```
package com.marakana.android.yamba;

import android.app.Activity;
import android.os.AsyncTask;
import android.os.Bundle;
import android.util.Log;
```

```java
import android.view.Menu;
import android.view.View;
import android.view.View.OnClickListener;
import android.widget.Button;
import android.widget.EditText;
import android.widget.Toast;

import com.marakana.android.yamba.clientlib.YambaClient;
import com.marakana.android.yamba.clientlib.YambaClientException;

public class StatusActivity extends Activity implements OnClickListener {
    private static final String TAG = "StatusActivity";
        private EditText editStatus;
        private Button buttonTweet;

        @Override
        protected void onCreate(Bundle savedInstanceState) {
                super.onCreate(savedInstanceState);
                setContentView(R.layout.activity_status);

                editStatus = (EditText) findViewById(R.id.editStatus);
                buttonTweet = (Button) findViewById(R.id.buttonTweet);

                buttonTweet.setOnClickListener(this);
        }

        @Override
        public void onClick(View view) {
                String status = editStatus.getText().toString();
                Log.d(TAG, "onClicked with status: " + status);

                new PostTask().execute(status); // ❶
        }

        private final class PostTask extends
                        AsyncTask<String, Void, String> { // ❷

                @Override
                protected String doInBackground(String... params) { // ❸
                        YambaClient yambaCloud =
                                new YambaClient("student", "password");
                        try {
                                yambaCloud.postStatus( params[0] ); // ❹
                                return "Successfully posted";
                        } catch (YambaClientException e) {
                                e.printStackTrace();
                                return "Failed to post to yamba service";
                        }
                }

                @Override
                protected void onPostExecute(String result) { // ❺
```

```
                    super.onPostExecute(result);

                    Toast.makeText(StatusActivity.this, result,
                                Toast.LENGTH_LONG).show(); // ➏
            }
        }

        @Override
        public boolean onCreateOptionsMenu(Menu menu) {
    // Inflate the menu; this adds items to the action bar if it is present.
                getMenuInflater().inflate(R.menu.status, menu);
                return true;
        }
}
```

➊ Once we have our AsyncTask set up, we can run it. To do so, we simply instantiate it and call execute() on it. The argument that we pass in is what goes into the doInBackground() call. Note that in this case we are passing a single string that is being converted into a string array in the actual method later on, which is an example of Java's variable number of arguments feature (*http://en.wikipedia.org/wiki/Variadic_function*). It is important to remember that this method executes on a separate thread, so you cannot update the UI from it.

➋ The PostTask class in this case is an inner class of StatusActivity. It also subclasses AsyncTask. Notice the use of Java generics (*http://en.wikipedia.org/wiki/Generics_in_Java*) to describe the data types that this AsyncTask will use in its methods. We'll explain these three types following this list. The first data type is used by doInBackground(), the second by onProgressUpdate(), and the third by onPostExecute().

➌ doInBackground() is the callback that specifies the actual work to be done on the separate thread, acting as if it's executing in the background. The argument String… is the first of the three data types that we defined in the list of generics for this PostTask inner class. The three dots indicate that this is an array of Strings, and you have to declare it that way, even though you want to pass only a single status.

➍ This is the call to the Yamba Client library that does all the magic of encoding it into a web service call. We're passing the first parameter of the input, which is the actual text of the status that comes in via the execute() call from on Click().

➎ onPostExecute() is called when our task completes. This is our callback method to update the user interface and tell the user that the task is done. It is important to know that this method executes on the application's main thread, which we also refer to as the UI thread.

❻   In this particular case, we are using a `Toast` feature of the Android UI to display a quick message on the screen. Notice that `Toast` uses the `makeText()` static method to make the actual message. Also, do not forget to include `show()`; otherwise, your message will never be displayed, and no error will be reported —a hard bug to find. The argument that this method gets is the value that `doInBackground()` returns, in this case a string. This also corresponds to the third generics data type in the `PostTask` class definition. The reference to `StatusActivity.this` represents the `Context` that our application is in. As a rule of thumb, whenever in an activity, pass that activity as the context object.

At this point, when the user clicks "Update Status," our activity creates a separate thread using `AsyncTask` and places the actual network operation on that thread. When done, the `AsyncTask` will update the main UI thread by popping up a `Toast` message to tell the user that the operation either succeeded or failed.

Other UI Events

So far, you have seen how to handle the click events by implementing `OnClickListener` and providing the `onClick()` method, which is invoked when the button is clicked. Imagine that we want to provide a little counter telling the user how many characters of input are still available out of the maximum of 140. The counter must change as the user is typing, without waiting for a click. To do that, we need another type of listener.

Android provides many different listeners for various events, such as touch and click. In this case, we're going to use `TextWatcher` to watch for text changes in the edit text field. Steps for this listener are similar to the steps for `OnClickListener` and many other listeners.

From the user's standpoint, we'll add another `TextView` to our layout to indicate how many characters are still available. This text will change color, from green to yellow to red, as the user approaches the 140-character limit.

In Java, we'll implement `TextWatcher` and attach it to the field where the user is typing the actual text. The `TextWatcher` methods will be invoked as the user changes the text, and based on the amount of text entered, we'll update the counter.

To start, we'll add another text view onto our layout. In Graphical Layout, we picked Small Text from the Form Widgets Palette, and dragged it to the top-left corner. We changed its ID to `textCount` and hardcoded the text to 140. Although hardcoding values is not a good practice because it precludes internationalization, in our case this value is just temporarily there to help us visually see what we're working on—we'll be updating it programmatically. See Example 7-7 for the source of the final result.

Example 7-7. The res/layout/status_activity.xml file

```xml
<RelativeLayout xmlns:android="http://schemas.android.com/apk/res/android"
    xmlns:tools="http://schemas.android.com/tools"
    android:layout_width="match_parent"
    android:layout_height="match_parent"
    android:paddingBottom="@dimen/activity_vertical_margin"
    android:paddingLeft="@dimen/activity_horizontal_margin"
    android:paddingRight="@dimen/activity_horizontal_margin"
    android:paddingTop="@dimen/activity_vertical_margin"
    tools:context=".StatusActivity" >

    <Button
        android:id="@+id/buttonTweet"
        android:layout_width="wrap_content"
        android:layout_height="wrap_content"
        android:layout_alignParentRight="true"
        android:layout_alignParentTop="true"
        android:layout_marginRight="17dp"
        android:text="@string/button_tweet" />

    <EditText
        android:id="@+id/editStatus"
        android:gravity="top"
        android:layout_width="match_parent"
        android:layout_height="match_parent"
        android:layout_alignParentLeft="true"
        android:layout_below="@+id/buttonTweet"
        android:layout_marginTop="10dp"
        android:ems="10"
        android:hint="@string/hint_status"
        android:inputType="textMultiLine" >

        <requestFocus />
    </EditText>

    <TextView //  ❶
        android:id="@+id/textCount"
        android:layout_width="wrap_content"
        android:layout_height="wrap_content"
        android:layout_alignLeft="@+id/editStatus"
        android:layout_alignTop="@+id/buttonTweet"
        android:text="140"
        android:textAppearance="?android:attr/textAppearanceSmall" />

</RelativeLayout>
```

❶ New TextView that represents how many characters are still available for the user to type. We start at 140 and then go down as the user enters text.

The version of `StatusActivity` shown in Example 7-8 implements the `TextWatcher` interface, and the new methods in this example appear at the end of the class. Initially the text of the counter is in green to indicate we can keep on typing. As we approach the maximum, the text turns yellow and eventually changes to red to indicate we are beyond the maximum message size.

Example 7-8. StatusActivity.java, final version

```java
package com.marakana.android.yamba;

import android.app.Activity;
import android.graphics.Color;
import android.os.AsyncTask;
import android.os.Bundle;
import android.text.Editable;
import android.text.TextWatcher;
import android.util.Log;
import android.view.Menu;
import android.view.View;
import android.view.View.OnClickListener;
import android.widget.Button;
import android.widget.EditText;
import android.widget.TextView;
import android.widget.Toast;

import com.marakana.android.yamba.clientlib.YambaClient;
import com.marakana.android.yamba.clientlib.YambaClientException;

public class StatusActivity extends Activity implements OnClickListener {
    private static final String TAG = "StatusActivity";
    private EditText editStatus;
    private Button buttonTweet;
    private TextView textCount; // ❶
    private int defaultTextColor; // ❷

    @Override
    protected void onCreate(Bundle savedInstanceState) {
        super.onCreate(savedInstanceState);
        setContentView(R.layout.activity_status);

        editStatus = (EditText) findViewById(R.id.editStatus);
        buttonTweet = (Button) findViewById(R.id.buttonTweet);
        textCount = (TextView) findViewById(R.id.textCount); // ❸

        buttonTweet.setOnClickListener(this);

        defaultTextColor =
                textCount.getTextColors().getDefaultColor(); // ❹
        editStatus.addTextChangedListener(new TextWatcher() { // ❺

            @Override
            public void afterTextChanged(Editable s) { // ❻
```

```
                          int count = 140 - editStatus.length(); // ❼
                          textCount.setText(Integer.toString(count));
                          textCount.setTextColor(Color.GREEN); // ❽
                          if (count < 10)
                            textCount.setTextColor(Color.RED);
                          else
                            textCount.setTextColor(defaultTextColor);
                    }

              @Override
              public void beforeTextChanged(CharSequence s,
                                                int start, int count,
                                                int after) { // ❾
              }

              @Override
              public void onTextChanged(CharSequence s,
                                                int start, int before,
                                                int count) { // ❿
              }

        });
  }

  @Override
  public void onClick(View view) {
        String status = editStatus.getText().toString();
        Log.d(TAG, "onClicked with status: " + status);

        new PostTask().execute(status);
  }

  private final class PostTask extends
                    AsyncTask<String, Void, String> {

        @Override
        protected String doInBackground(String... params) {
              YambaClient yambaCloud =
                    new YambaClient("student", "password");
              try {
                    yambaCloud.postStatus(params[0]);
                    return "Successfully posted";
              } catch (YambaClientException e) {
                    e.printStackTrace();
                    return "Failed to post to yamba service";
              }
        }

        @Override
        protected void onPostExecute(String result) {
              super.onPostExecute(result);
```

```
                    Toast.makeText(StatusActivity.this, result,
                                Toast.LENGTH_LONG).show();
        }
    }

    @Override
    public boolean onCreateOptionsMenu(Menu menu) {
// Inflate the menu; this adds items to the action bar if it is present.
            getMenuInflater().inflate(R.menu.status, menu);
            return true;
    }
}
```

❶ textCount is our text view, defined in Example 7-7.

❷ This variable will hold the default text color. Because we don't now what theme
    the application may use down the road, we don't want to make any assumptions
    about the default color of the text. Rather, we'll calculcate it at runtime.

❸ First, we need to find the textCount in the inflated layout.

❹ This is where we determine the default color of the text.

❺ We attach the TextWatcher listener to the status text area. Unlike in onClick
    Listener, we chose to implement the TextWatcher as an anonymous inner class
    (*http://en.wikipedia.org/wiki/Anonymous_class#Unnamed*). This is a very
    standard practice in Android, especially for the UI event handlers.

❻ TextWatcher has a number of callbacks, but we use only afterTextCh
    anged(). This method is called once the user makes a change in the actual text
    field. Here, we set the initial text to 140 because that's the maximum length of a
    status message in our app. Note that TextView takes text as a value, so we convert
    a number to text here.

❼ Here we do some math to figure out how many characters are left, given the 140-
    character limit.

❽ The textCount field will change color dynamically based on the number of
    remaining characters. In this case, we start with the defaultTextColor, and
    switch to java.awt.Color.RED once the user has fewer than 10 characters left.
    Notice that the Color class is part of the Android framework and not Java. In
    other words, we're using android.graphics.Color and not java.awt.Color.
    Color.RED is one of the few colors defined as a constant in this class (more on
    colors in the next section). Next, based on the availability of the text, we update
    the color of the counter. So, if more than 10 characters are available, we are still
    in the green. Fewer than 10 means we are approaching the limit, thus the counter
    turns yellow. If we are past the limit of 140 characters, the counter turns red.

❾ This method is called just before the actual text replacement is completed. In this case, we don't need this method, but as part of implementing the `TextWatch er` interface, we must provide its implementation, even though it's empty.

❿ Similarly, we are not using `onTextChanged()` in this case, but must provide its blank implementation. Figure 7-19 shows what the `TextWatcher` looks like in our application when running.

Figure 7-19. StatusActivity with text counter

Alternative Resources

Android supports multiple competing sets of resources. For example, you could have multiple versions of a *strings.xml* file, *activity_status.xml* layout, or other resources. You might want multiple versions of the same resource so that the best version can be used under different circumstances. We touched on this in "Drawable Resources" on page 56.

Imagine that your application is used in another country with a different language. In that case, you could provide a *strings.xml* version specifically for that language. Or imagine that a user runs your application on a different device, with a different screen that has more pixels. In that case, you'd want versions of your images specifically for this screen's pixel density. Similarly, users might simply rotate the device from portrait to landscape mode. Our application will redraw properly, but there are further enhancements we could make to the layout of the UI given the orientation of the screen.

Android provides for all these cases in an elegant way. Basically, you simply need to create alternative folders for specific constraints. For example, our standard layout files go into the */res/layout* folder, but if we wanted to provide an alternative layout specifically for landscape mode, we'd simply create a new file called */res/layout-land/activity_status.xml*. And if we wanted to provide a translated version of our *strings.xml* file for users who are in a French-speaking part of Canada, we'd put it in file called *res/values-fr-rCA/strings.xml*.

As you see from these examples, alternative resources work by specifying the qualifiers in the names of their resource folders. In the case of the French Canadian strings, Android knows that the first qualifier -fr refers to language, and the second qualifier -rCA specifies that the region is Canada. In both cases, we use two-letter ISO codes (*http://www.loc.gov/standards/iso639-2/php/code_list.php*) to specify the country. So in this case, if the user is in Quebec and her device is configured to favor the French language, Android will look for string resources in the */res/values-fr-rCA/strings.xml file*. If it doesn't find a specific resource, it will fall back to the default */res/values/strings.xml* file. Also, if the user is in France, Android will use the default resource, because our French Canadian qualifiers do not match French for France.

Using qualifiers, you can create alternative resources for languages and regions, screen sizes and orientations, device input modes (touch screen, stylus), keyboard or no keyboard, and so on. But how do you figure out this naming convention for resource folder names?

Let's create an alternative resource for the landscape view of our screen. The easiest solution is to use Eclipse's New Android XML File dialog (see Figure 7-20). To open the New Android XML File dialog, choose File → New → Android XML File from the Eclipse menu. Choose Layout as the resource type, and name it the same as our layout filename, i.e., *activity_status.xml*. Click Next, and on the next screen choose Orientation → Landscape. Click Finish. This will create a new file under *res/layout-land/activity_status.xml*.

The easiest way to create our landscape layout is to start with the default one that we have already created. To do that, copy and paste the XML from *res/layout/activity_status.xml* to *res/layout-land/activity_status.xml*. Then, customize the user interface for that orientation, for example, moving the button and the counter to the right and giving the text area a bit more vertical space so that the keyboard doesn't obstruct it (Figure 7-21).

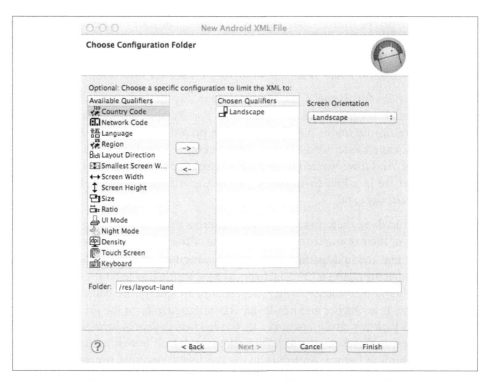

Figure 7-20. Alternative resources with the New Android XML File dialog

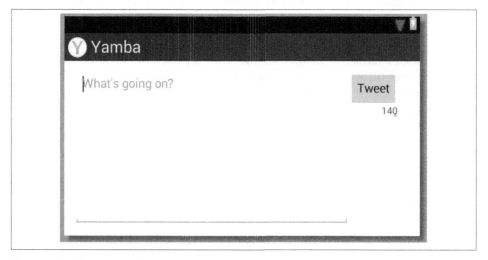

Figure 7-21. Landscape layout

You can now run your application and try to rotate the screen in order to see how the system will render the appropriate resource file. On the emulator, use Crtl-F11.

Summary

By the end of this chapter, your application should run and should look like Figure 7-22. It should also successfully post your tweets to your Twitter account. You can verify it is working by logging in to an online service of your choice that supports the Twitter API, such as *http://yamba.marakana.com*, using the same username and password that are hardcoded in the application.

Figure 7-22. StatusActivity

Figure 7-23 illustrates what we have done so far as part of the design outlined in Figure 6-4.

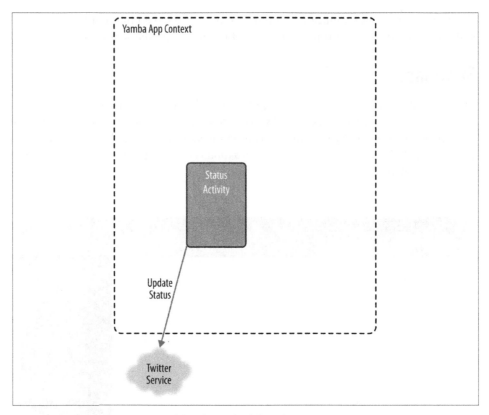

Figure 7-23. Yamba, as created by the end of this chapter

Fragments

In Android 3.0 (API Level 11), Android introduced the Fragments API. This was in response to a growing need to accommodate multiple screen sizes (such as tablets versus phones) and orientations (landscape versus portrait). To do this, it was necessary to modularize the views (the UI) such that it would be easy to separate the Activity container from the UI. This enables the developer to create a more responsive and easy-to-build interface to the user's needs—such as changing the interface on the fly rather than having to create completely new containers for every configuration.

Fragment Example

To show how easy it is to convert an activity into a fragment, let's start with the `Status Activity` we created in a previous chapter. For reference, Example 8-1 is a copy of the old *StatusActivity.java* file and Example 8-2 is a copy of the layout XML (the *activity_status.xml* file) for `StatusActivity`.

Example 8-1. Old StatusActivity

```
package com.marakana.android.yamba;

import android.app.Activity;
import android.graphics.Color;
import android.os.AsyncTask;
import android.os.Bundle;
import android.text.Editable;
import android.text.TextWatcher;
import android.util.Log;
import android.view.Menu;
import android.view.View;
import android.view.View.OnClickListener;
import android.widget.Button;
import android.widget.EditText;
import android.widget.TextView;
```

```
import android.widget.Toast;

import com.marakana.android.yamba.clientlib.YambaClient;
import com.marakana.android.yamba.clientlib.YambaClientException;

public class StatusActivity extends Activity implements OnClickListener {
        private static final String TAG = "StatusActivity";
        private EditText editStatus;
        private Button buttonTweet;
        private TextView textCount;
        private int defaultTextColor;

        @Override
        protected void onCreate(Bundle savedInstanceState) {
                super.onCreate(savedInstanceState);
                setContentView(R.layout.activity_status);

                editStatus = (EditText) findViewById(R.id.editStatus);
                buttonTweet = (Button) findViewById(R.id.buttonTweet);
                textCount = (TextView) findViewById(R.id.textCount);

                buttonTweet.setOnClickListener(this);

                defaultTextColor = textCount.getTextColors().getDefaultColor();
                editStatus.addTextChangedListener(new TextWatcher() {

                        @Override
                        public void afterTextChanged(Editable s) {
                                int count = 140 - editStatus.length();
                                textCount.setText(Integer.toString(count));
                                textCount.setTextColor(Color.GREEN);
                                if (count < 10)
                                  textCount.setTextColor(Color.RED);
                                else
                                  textCount.setTextColor(defaultTextColor);
                        }

                        @Override
                        public void beforeTextChanged(CharSequence s,
                                                        int start, int count,
                                                        int after) {
                        }

                        @Override
                        public void onTextChanged(CharSequence s,
                                                        int start, int before,
                                                        int count) {
                        }

                });
        }
```

```java
    @Override
    public void onClick(View view) {
            String status = editStatus.getText().toString();
            Log.d(TAG, "onClicked with status: " + status);

            new PostTask().execute(status);
    }

    private final class PostTask extends AsyncTask<String, Void, String> {

            @Override
            protected String doInBackground(String... params) {
                    YambaClient yambaCloud =
                            new YambaClient("student", "password");
                    try {
                            yambaCloud.postStatus(params[0]);
                            return "Successfully posted";
                    } catch (YambaClientException e) {
                            e.printStackTrace();
                            return "Failed to post to yamba service";
                    }
            }

            @Override
            protected void onPostExecute(String result) {
                    super.onPostExecute(result);

                    Toast.makeText(StatusActivity.this, result,
                                    Toast.LENGTH_LONG).show();
            }
    }

    @Override
    public boolean onCreateOptionsMenu(Menu menu) {
// Inflate the menu; this adds items to the action bar if it is present.
            getMenuInflater().inflate(R.menu.status, menu);
            return true;
    }

}
```

Example 8-2. Old StatusActivity layout: activity_status.xml

```xml
<RelativeLayout xmlns:android="http://schemas.android.com/apk/res/android"
    xmlns:tools="http://schemas.android.com/tools"
    android:layout_width="match_parent"
    android:layout_height="match_parent"
    android:paddingBottom="@dimen/activity_vertical_margin"
    android:paddingLeft="@dimen/activity_horizontal_margin"
    android:paddingRight="@dimen/activity_horizontal_margin"
    android:paddingTop="@dimen/activity_vertical_margin"
    tools:context=".StatusActivity" >
```

```
    <Button
        android:id="@+id/buttonTweet"
        android:layout_width="wrap_content"
        android:layout_height="wrap_content"
        android:layout_alignParentRight="true"
        android:layout_alignParentTop="true"
        android:layout_marginRight="17dp"
        android:text="@string/button_tweet" />

    <EditText
        android:id="@+id/editStatus"
        android:gravity="top"
        android:layout_width="match_parent"
        android:layout_height="match_parent"
        android:layout_alignParentLeft="true"
        android:layout_below="@+id/buttonTweet"
        android:layout_marginTop="10dp"
        android:ems="10"
        android:hint="@string/hint_status"
        android:inputType="textMultiLine" >

        <requestFocus />
    </EditText>

    <TextView
        android:id="@+id/textCount"
        android:layout_width="wrap_content"
        android:layout_height="wrap_content"
        android:layout_alignLeft="@+id/editStatus"
        android:layout_alignTop="@+id/buttonTweet"
        android:text="140"
        android:textAppearance="?android:attr/textAppearanceSmall" />

</RelativeLayout>
```

To switch to the Fragment API, we need to move the StatusActivity logic into a Fragment extended class. Example 8-3 creates a StatusFragment that extends Fragment and continues to implement OnClickListener and TextWatcher just like the StatusActivity. If you carefully compare the two examples, you will notice that they are extremely similar. This shows just how simple the switch is.

Example 8-3. StatusFragment

```
package com.marakana.android.yamba;

import android.app.Fragment;
import android.graphics.Color;
import android.os.AsyncTask;
import android.os.Bundle;
import android.text.Editable;
import android.text.TextWatcher;
import android.util.Log;
```

```java
import android.view.LayoutInflater;
import android.view.View;
import android.view.View.OnClickListener;
import android.view.ViewGroup;
import android.widget.Button;
import android.widget.EditText;
import android.widget.TextView;
import android.widget.Toast;

import com.marakana.android.yamba.clientlib.YambaClient;
import com.marakana.android.yamba.clientlib.YambaClientException;

public class StatusFragment extends Fragment implements OnClickListener {
    private static final String TAG = "StatusFragment";
    private EditText editStatus;
    private Button buttonTweet;
    private TextView textCount;
    private int defaultTextColor;

    @Override
    public View onCreateView(LayoutInflater inflater, ViewGroup container,
                Bundle savedInstanceState) {

        View view = inflater
                .inflate(R.layout.fragment_status, container, false);

        editStatus = (EditText) view.findViewById(R.id.editStatus);
        buttonTweet = (Button) view.findViewById(R.id.buttonTweet);
        textCount = (TextView) view.findViewById(R.id.textCount);

        buttonTweet.setOnClickListener(this);

        defaultTextColor = textCount.getTextColors().getDefaultColor();
        editStatus.addTextChangedListener(new TextWatcher() {

            @Override
            public void afterTextChanged(Editable s) {
                int count = 140 - editStatus.length();
                textCount.setText(Integer.toString(count));
                textCount.setTextColor(Color.GREEN);
                if (count < 10)
                    textCount.setTextColor(Color.RED);
                else
                    textCount.setTextColor(defaultTextColor);
            }

            @Override
            public void beforeTextChanged(CharSequence s,
                                    int start, int count,
                                    int after) {
            }
```

```java
        @Override
        public void onTextChanged(CharSequence s,
                                            int start, int before,
                                            int count) {
        }

    });

    return view;
}


@Override
public void onClick(View view) {
    String status = editStatus.getText().toString();
    Log.d(TAG, "onClicked with status: " + status);

    new PostTask().execute(status);
}

private final class PostTask extends AsyncTask<String, Void, String> {

    @Override
    protected String doInBackground(String... params) {
        YambaClient yambaCloud =
                new YambaClient("student", "password");
        try {
                yambaCloud.postStatus(params[0]);
                return "Successfully posted";
        } catch (YambaClientException e) {
                e.printStackTrace();
                return "Failed to post to yamba service";
        }
    }

    @Override
    protected void onPostExecute(String result) {
        super.onPostExecute(result);

        Toast.makeText(StatusFragment.this.getActivity(),
                        result, Toast.LENGTH_LONG).show();
    }
}
}
```

To add Fragments, we had to import three additional classes: `android.app.Fragment`, `android.view.LayoutInflater`, and `android.view.ViewGroup`. Our `StatusFrag ment` class overrides the `onCreateView` method, and creates a view there that performs the activities done by `onCreate` in the original `StatusActivity`. The rest of the code is the same as before. The following line in the `onCreateView()` method:

```
View view = inflater.inflate(R.layout.fragment_status, container, false);
```

uses the inflater from `LayoutInflater` to establish a view for the fragment.

The layout XML for the this new fragment is also simple. Rename the *activity_status.xml* file that is located in *res/layout* to *fragment_status.xml* (keeping it in *res/ layout*). This is so we are clear what the layout file is for. (Technically, we could have reused the *activity_status.xml* file as is by referring to it in the inflater logic as `R.lay out.status`.)

Now that the fragment is out of the way, we can change the `StatusActivity` (Example 8-4) to reflect the new setup. This change in fact simplifies the `StatusActiv ity` greatly, because all we do now within the activity is specify a new content view layout.

Example 8-4. New StatusActivity

```
package com.marakana.android.yamba;

import android.app.Activity;
import android.os.Bundle;
import android.view.Menu;

public class StatusActivity extends Activity {

        @Override
        protected void onCreate(Bundle savedInstanceState) {
                super.onCreate(savedInstanceState);
                setContentView(R.layout.new_activity_status);
        }

        @Override
        public boolean onCreateOptionsMenu(Menu menu) {
    // Inflate the menu; this adds items to the action bar if it is present.
                getMenuInflater().inflate(R.menu.status, menu);
                return true;
        }
}
```

It is this new layout (shown in Example 8-5) that supplies the hook to the newly created fragment. You should put the code for the layout in a file called *new_activity_status.xml* in the *res/layout* directory, As you can see, a new `<fragment />` tag is placed there, referencing the new fragment class: `com.marakana.android.yamba.Status Fragment`.

Example 8-5. StatusActivity layout: new_activity_status.xml

```xml
<?xml version="1.0" encoding="utf-8"?>
<FrameLayout xmlns:android="http://schemas.android.com/apk/res/android"
    android:layout_width="match_parent"
    android:layout_height="match_parent" >

    <fragment
        android:id="@+id/fragment_status"
        android:name="com.marakana.android.yamba.StatusFragment"
        android:layout_width="match_parent"
        android:layout_height="match_parent" />

</FrameLayout>
```

Fragment Life Cyle

The fragment life cycle is extremely important to understand, particularly how and when methods in the fragment are called during the activity's life cycle. Fragments go through a life cycle similar to activities: they are created, started and stopped, resumed and paused, and finally destroyed. As with an activity, you should save to disk or to a database any data in the fragment you want persisted in your onPause method for the fragment. Figure 8-1 shows the fragment methods in parallel to the activity's states. The important methods to note are the following:

onCreateView()
> Called when the view is created

onResume()
> Called when the activity's onResume() is called

onPause()
> Parallel's the activity's onPause()

onDestroyView()
> Called when the view is destroyed

This is why, in our example, we inflated the layout in onCreateView() so that when the activity is created, the view is created accordingly.

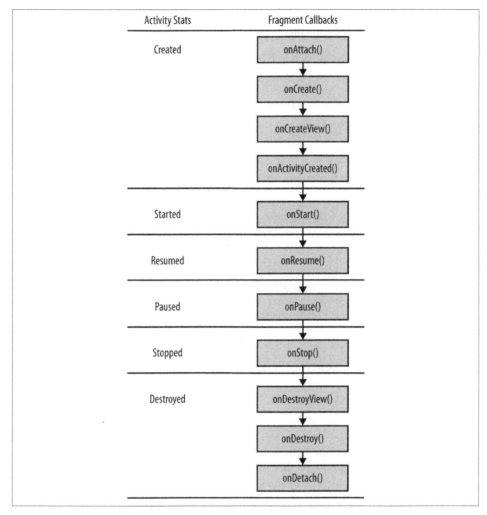

Activity Stats	Fragment Callbacks
Created	onAttach()
	onCreate()
	onCreateView()
	onActivityCreated()
Started	onStart()
Resumed	onResume()
Paused	onPause()
Stopped	onStop()
Destroyed	onDestroyView()
	onDestroy()
	onDetach()

Figure 8-1. Activity and fragment life cycle

Dynamically Adding Fragments

In Example 8-5, we have added our status fragment to the activity by creating an XML layout file that our activity inflated, at which time the fragment was also loaded and created. Because initialization of the fragment happens in an XML file, this is called *static initialization*. At a closer look, you may notice that most of the XML file itself is almost just noise—it's there just to define that we need to load a fragment.

As you know from before, XML ultimately always becomes Java, so everything that can be done statically can also be done dynamically. Sometimes one approach is cleaner than others. Just for comparison purposes, let's refactor our code to attach status fragment to status activity dynamically.

Our status activity can now be seen in Example 8-6, which shows how to load a fragment dynamically (in other words, without the need for an XML file).

Example 8-6. StatusActivity, refactored

```
package com.marakana.android.yamba;

import android.os.Bundle;

public class StatusActivity extends Activity {

    @Override
    protected void onCreate(Bundle savedInstanceState) {
        super.onCreate(savedInstanceState);

        // Check if this activity was created before
        if (savedInstanceState == null) { // ❶
            // Create a fragment
            StatusFragment fragment = new StatusFragment(); // ❷
            getFragmentManager()
                        .beginTransaction()
                        .add(android.R.id.content, fragment,
                                fragment.getClass().getSimpleName())
            .commit(); // ❸
        }
    }

}
```

❶ We no longer call setContentView(). Instead, we check whether this is the first time this activity is created, because we could also have gotten to this point in the code when the activity exists and the screen has been rotated by the user.

❷ First time around, we need to instantiate the status fragment.

❸ Next, in one go, we obtain the fragment manager from the current context, start a transaction, attach this new fragment to the root of this activity identified by the system ID android.R.id.content, and commit this transaction. That's it.

We no longer need *res/layout/activity_status.xml*—you can safely delete this file because it's never read.

At this point, the app looks like Figure 8-2.

Figure 8-2. Status activity

Summary

In this chapter, we covered the basics of the Fragments API by converting one of the activities we created in a prior chapter. We also explained the modularity that the Fragments API provides.

Figure 8-3 illustrates what we have done so far as part of the design outlined earlier in Figure 6-4.

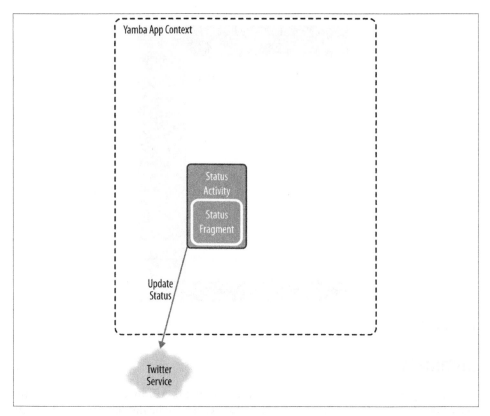

Figure 8-3. Yamba completion

Intents, Action Bar, and More

In this chapter, you will learn how to create preferences for your application, how the filesystem is organized, and how to use intents and the options menu to jump from one activity to another.

Preferences

Preferences are user-specific settings for an application. They usually consist of some configuration data as well as a user interface to manipulate that data.

In the user interface, preferences can be simple text values, checkboxes, selections from a pull-down menu, or similar items. From a data point of view, preferences are a collection of name-value pairs (*http://en.wikipedia.org/wiki/Attribute-value_pair*), also known as key-value or attribute-value pairs. The values are basic data types, such as integers, booleans, and strings.

Our micro-blogging application needs to connect to a specific server in the cloud using specific user account information. For that, Yamba needs to know the username and password for that account as well as the URL of the server it's connecting to. This URL is also known as the API root. So our interface will have three fields where the user can enter and edit his username, his password, and the API root. This data will be stored as strings.

To enable our app to handle user-specific preferences, we need to build a screen to enter the information, Java code to validate and process that information, and some kind of mechanism to store the information.

All this sounds like a lot of work, but Android provides a framework to help streamline working with user preferences. First, we'll define what our preference data looks like in a preference resource file.

To create preferences for our application, we need to:

1. Create a Preference resource file called *settings.xml*.
2. Implement the *SettingsActivity.java* file that inflates that resource file.
3. Register this new activity with the *AndroidManifest.xml* file.
4. Provide a way to start that activity from the rest of the application.

Preference Resource

We are going to start by creating *settings.xml*, a resource file that outlines what our preference screen will look like. The easiest way to create it is to use the New Android XML File tool in Eclipse, as shown in Figure 9-1. To start the New Android XML File dialog, go to File → New → Android XML File.

The key is to give the new file a name, in this case *settings.xml*, and to choose Preference for the type of resource. The tool should automatically suggest creating this new file in the */res/xml* folder and that the root element for the XML file should be `Preferen ceScreen`. As discussed in "Alternative Resources" on page 124, we could create alternative versions of this same resource by applying various qualifiers, such as screen size and orientation, language and region, etc.

When you click Finish, Eclipse will create a new file for you and open it. Eclipse typically opens XML files that it has some specific knowledge about in a view that lets you easily manipulate the content—a developer-friendly view.

In this view, you can create the username preference entry by selecting PreferenceScreen on the left, and then choosing Add → EditTextPreference. On the right side, expand the "Attributes from Preferences" section. Eclipse will offer you a number of attributes to set for this `EditTextPreference`.

Not all attributes are equally important. Typically, you will care about the following:

Key
> A unique identifier for each preference item. This is how we'll look up a particular preference later.

Title
> The preference name that the user will see. It should be a short name that fits on a single line of the preference screen.

Summary
> A short description of this preference item. This is optional, but highly recommended so you and other people can understand later what you've created.

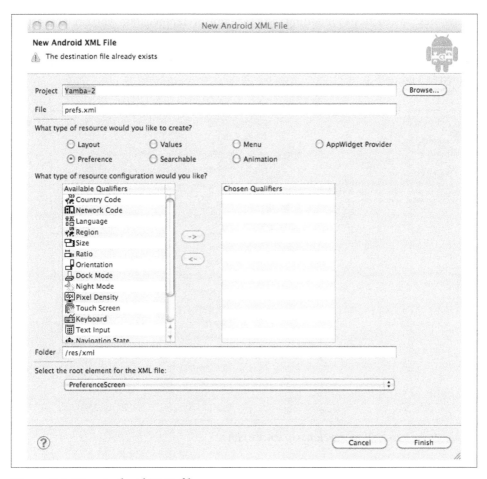

Figure 9-1. New Android XML file

For the username preference, we'll put "username" for its key. We will define the Title and Summary in *strings.xml*, because this is the best practice.

Instead of modifying the *strings.xml* file directly, you can use an Eclipse shortcut. Here's how it goes:

1. Click Browse and select New String. This will open a dialog to create a new string resource.

2. Enter **username** for the R.string. value and **Username** for the String value.

3. Click OK. Eclipse will insert a new string resource in *strings.xml*.

You can now pick that value from the list of resources.

Using these instructions for adding the Username preference item, you can now repeat the same steps for the Password and API Root items.

You can switch to the actual XML code by clicking the tab at the bottom of the window, shown in Figure 9-2.

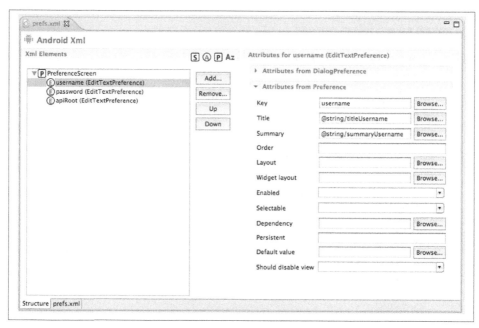

Figure 9-2. Prefs.xml in developer-friendly view

The raw XML for the preference resource looks like the code shown in Example 9-1.

Example 9-1. The res/xml/settings.xml file

```xml
<?xml version="1.0" encoding="utf-8"?>
<PreferenceScreen xmlns:android="http://schemas.android.com/apk/res/android" >

    <EditTextPreference
        android:key="username"
        android:summary="@string/username_summary"
        android:title="@string/username" />
    <EditTextPreference
        android:inputType="textPassword"
        android:title="@string/password"
        android:key="password" android:summary="@string/password_summary"/>
</PreferenceScreen>
```

`<PreferenceScreen>` is the root element that defines our main preference screen. It has three children, all of the `EditTextPreference` type. This widget is simply a piece of editable text. Other elements commonly used to enter preferences are `<CheckBoxPre ference>`, `<ListPreference>`, and so on.

The main property of any of these elements is the key. The key is how we'll look up these values later on. Remember, preferences are just a set of key-value pairs at the end of the day.

As we said earlier, although Eclipse does provide developer-friendly tools to manage XML files, you often run into certain limitations with Eclipse. For example, we would like to hide the actual text that the user types in the password field, which is a common practice. Android provides support for that, but Eclipse tools haven't yet integrated this function. Because we can always edit the XML directly, in this case we add an `an droid:inputType="textPassword"` property to our password property. This will cause the password to be masked while the user types it in.

SettingsActivity

Now that we have the preferences defined in their own XML resource file, we can create the activity to display these preferences. You may recall from "Activities" on page 64 that every screen in an Android app is an activity. So, to display the screen where a user enters the username and password for his online account, we'll create an activity to handle that screen. This will be a special preference-aware activity.

To create an activity, we create a new Java class. In Eclipse, select your package under your *src* folder, right-click the package, and select New→Class. A New Java Class window will pop up. Enter **SettingsActivity** for the Name and click Finish. This will create a *SettingsActivity.java* file under your package in your source folder. Then do the same thing and create a `SettingsActivity` class.

Our `SettingsActivity` class, shown in Example 9-2, is a very simple Java file. This is because we inherit from `Activity` and use the `SettingsFragment` class shown in Example 9-3. The `SettingsFragment` class extends the `SettingsActivity` class, an Android framework class that knows how to handle preferences.

Example 9-2. SettingsActivity.java

```
package com.marakana.android.yamba;

import android.os.Bundle;

public class SettingsActivity extends Activity {

    @Override
    protected void onCreate(Bundle savedInstanceState) {
            super.onCreate(savedInstanceState);

            // Check whether this activity was created before
            if (savedInstanceState == null) {
```

```
                        // Create a fragment
                        SettingsFragment fragment = new SettingsFragment(); // ❶
                        getFragmentManager()
                                        .beginTransaction()
                                        .add(android.R.id.content, fragment,
                                                        fragment.getClass().getSimpleName())
                        .commit(); // ❷
                }
        };
}
```

❶   As before, in case this is the first time we're creating this activity, we create the instance of the fragment that will be housed here.

❷   Next, we obtain the fragment transaction from the fragment manager, and add this fragment to the activity's main content.

Next, we get to create the actual settings fragment.

Example 9-3. SettingsFragment.java

```
package com.marakana.android.yamba;

import android.content.Intent;
import android.content.SharedPreferences;
import android.content.SharedPreferences.OnSharedPreferenceChangeListener;
import android.os.Bundle;
import android.preference.PreferenceFragment;
import android.preference.PreferenceManager;

public class SettingsFragment extends PreferenceFragment implements
        OnSharedPreferenceChangeListener { // ❶
        private SharedPreferences prefs;

        @Override
        public void onCreate(Bundle savedInstanceState) { // ❷
                super.onCreate(savedInstanceState);
                addPreferencesFromResource(R.xml.settings); // ❸
        }

        @Override
        public void onStart() {
                super.onStart();
                prefs = PreferenceManager.getDefaultSharedPreferences(getActivity());
                prefs.registerOnSharedPreferenceChangeListener(this);
        }

}
```

❶   Unlike regular fragments, SettingsFragment will subclass (i.e., extend) the PreferenceFragment class.

❷ Just like any other fragment, we override the `onCreate()` method to initialize the fragment.

❸ Our preference fragment sets its content from the *settings.xml* file via a call to `addPreferencesFromResource()`.

Update the Manifest File

Whenever we create one of these main building blocks (activities, services, broadcast receivers, or content providers), we need to define them in the *AndroidManifest.xml* file. In this case, we have a new `SettingsActivity` and must add it to the manifest file.

Just as with any Android XML file, opening *AndroidManifest.xml* in Eclipse typically will bring up the developer-friendly view of that file. In this file view, you could choose the Application tab, and then under Application Nodes, choose Add → Activity and name it `.SettingsActivity`.

However, we can also do this straight from the raw XML by clicking the AndroidManifest.xml tab on the bottom of this window. We find Eclipse to be useful for the initial creation of XML files, but after that, editing the raw XML is often faster and gives you much more control.

 When editing code in Eclipse, you can use the Ctrl-spacebar key shortcut to invoke the type-ahead feature of Eclipse. This is very useful for both XML and Java code and is context-sensitive, meaning Eclipse is smart enough to know what could possibly be entered at that point in the code. Using Ctrl-spacebar makes your life as a programmer much easier because you don't have to remember long method names and tags, and it helps avoid typos.

So our manifest file now looks like the code shown in Example 9-4.

Example 9-4. AndroidManifest.xml

```
<?xml version="1.0" encoding="utf-8"?>
<manifest xmlns:android="http://schemas.android.com/apk/res/android"
    package="com.marakana.android.yamba"
    android:versionCode="1"
    android:versionName="1.0" >

    <uses-sdk
        android:minSdkVersion="11"
        android:targetSdkVersion="17" />

    <uses-permission android:name="android.permission.INTERNET" />

    <application
```

```
        android:allowBackup="true"
        android:icon="@drawable/ic_launcher"
        android:label="@string/app_name"
        android:theme="@style/AppTheme" >
        <activity
            android:name="com.marakana.android.yamba.StatusActivity"
            android:label="@string/status_update" >
            <intent-filter>
                <action android:name="android.intent.action.MAIN" />

                <category android:name="android.intent.category.LAUNCHER" />
            </intent-filter>
        </activity>

        <!-- ❶ -->
        <activity
            android:name="com.marakana.android.yamba.SettingsActivity"
            android:label="@string/action_settings" >
        </activity>
    </application>

</manifest>
```

❶   Defines the new `SettingsActivity`.

We now have a new preference activity, but there's no good way for the user to get at it yet and express preferences. We need a way to launch this new activity. For that, we use the options menu.

The Action Bar

The options menu is an Android user interface component that provides standardized menus to applications. Pre-Honeycomb, options menus appear at the bottom of the screen when the user presses the Menu button on the device. Now, options menu has been replaced by the Action Bar. From a developer's point of view, the Action Bar is the same as the older system, with couple of additional features. From a user's point of view, though, the Action Bar shows at the top of the application, as a navigation bar would on a web page.

To add support for the options menu to an application, we need to do the following:

1. Create the *main.xml* resource where we specify what the menu consists of.

2. Add `onCreateOptionsMenu()` to the activity that should have this menu. This is where we inflate the *menu.xml* resource.

3. Provide handling of menu events in `onOptionsItemSelected()`.

Creating a Blank Main Activity

Status activity is the view from which we tweet. But, that's likely not going to be the entry point into the application. After all, the user much more frequently consumes the timeline than tweets status updates. So, we're going to introduce a main entry point into our app, the Main Activity. For now, this activity will be just an empty screen, but it will also load our menus.

So go ahead and create a new blank activity using File → New → Other → Android Activity and call it *MainActivity*:

Eclipse ADT will register your new activity in the manifest file, but you may have to manually designate `MainActivity` to actually be the main entry point into the app. To do that, you need to move the intent filter that we had on the Status Activity over to Main Activity. Your new code would look like Example 9-5.

Example 9-5. AndroidManifest.xml with Main Activity

```
<?xml version="1.0" encoding="utf-8"?>
<manifest xmlns:android="http://schemas.android.com/apk/res/android"
    package="com.marakana.android.yamba"
    android:versionCode="1"
    android:versionName="1.0" >

    <uses-sdk
        android:minSdkVersion="11"
        android:targetSdkVersion="17" />
```

```
    <uses-permission android:name="android.permission.INTERNET" />

    <application
        android:allowBackup="true"
        android:icon="@drawable/ic_launcher"
        android:label="@string/app_name"
        android:theme="@style/AppTheme" >
        <activity
            android:name="com.marakana.android.yamba.StatusActivity"
            android:label="@string/status_update" >
        </activity>
        <activity android:name="com.marakana.android.yamba.MainActivity" >
            <!-- ❶ -->
            <intent-filter>
                <action android:name="android.intent.action.MAIN" />

                <category android:name="android.intent.category.LAUNCHER" />
            </intent-filter>
        </activity>
        <activity
            android:name="com.marakana.android.yamba.SettingsActivity"
            android:label="@string/action_settings" >
        </activity>
    </application>

</manifest>
```

❶ This particular intent filter makes this particular activity the one that opens when
   the user starts the app. By moving these four lines from Status Activity over to
   Main Activity, we made Main Activity the *home page* of our app.

The Menu Resource

Once you create the Main Activity using Eclipse, the ADT plug-in will create a new
folder called */res/menu* that contains the *main.xml* file and will open this file in the
developer-friendly view (see Figure 9-3).

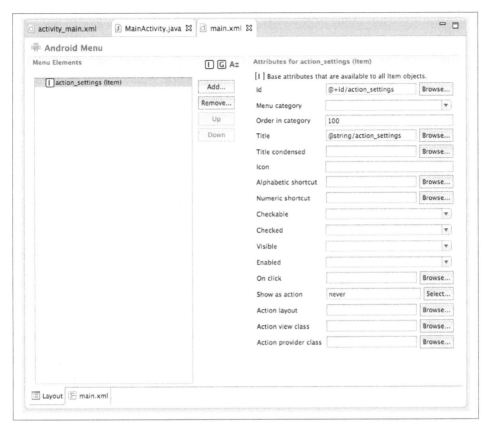

Figure 9-3. Menu.xml in developer-friendly view

In this view, you can click Add → Item, which will add a new menu item to your menu. In the Attributes section on the right, you can see more than a dozen attributes that we can set for this menu item. Just as before, not all attributes are equally important. We'll make sure to set these:

Id

The unique identifier of this resource. Just as when we designed the layout in Chapter 7, this identifier is typically of the form @+id/*someId*, where *someId* is the name that you give it. This name should contain only letters, numbers, and the underscore character.

Title

The title of this menu as it will appear on the display. Keep in mind that screen space typically is limited, so keep the title short. Additionally, you can provide a "Title condensed" attribute to specify a shorter version of the title that will be shown instead if space is limited. Just like before, best practice is to define the actual text value of the title in the *strings.xml* resource and just refer to the defined text value here.

Icon

The icon that displays along with the menu item's title. Although not required, it is a very useful visual cue from a usability point of view. In this case it also illustrates how to point to Android system resources.

The next section describes these resources in more detail.

Android System Resources

Just as your application can have resources, so can the Android system. Like most other operating systems, Android comes with some preloaded images, graphics, sound clips, and other types of resources. Recall that our app resources are in */res*. To refer to Android system resources, prefix them with the `android:` keyword in XML; for example, `@an droid:drawable/ic_menu_preferences`. If you are referring to an Android system resource from Java, use `android.R` instead of the usual `R` reference.

 The actual resource files are in your SDK, inside a specific platform folder. For example, if you are using Android 9 (Gingerbread), the resource folder would be *android-sdk/platforms/android-9/data/res/*.

Now, we're going to create a menu resource with a couple of buttons that our application will use, such as *Tweet*, *Settings*, *Refresh*, and *Purge*. The raw XML of *main.xml* is shown in Example 9-6.

Example 9-6. The res/menu/main.xml file

```
<menu xmlns:android="http://schemas.android.com/apk/res/android" >

    <item
        android:id="@+id/action_settings"
        android:orderInCategory="100"
        android:showAsAction="never"
        android:title="@string/action_settings"/>
    <item
        android:id="@+id/action_tweet"
        android:icon="@android:drawable/ic_menu_add"
        android:showAsAction="always|withText"
        android:title="@string/tweet"/>
    <item
        android:id="@+id/action_refresh"
        android:icon="@android:drawable/ic_menu_rotate"
        android:showAsAction="always"
        android:title="@string/refresh"/>
```

```
<item
    android:id="@+id/action_purge"
    android:icon="@android:drawable/ic_menu_delete"
    android:title="@string/purge">
</item>

</menu>
```

Next, we need to display this menu.

Loading the Menu

Recall that the options menu is loaded by your activity when the user clicks her device's Menu button. The first time the Menu button is pressed, the system calls the activity's onCreateOptionsMenu() method to inflate the menu from the *menu.xml* resource. This process is similar to inflating the user interface from layout resources, discussed in "The StatusActivity Java Class" on page 104. Basically, the *inflater* reads the XML code, creates a corresponding Java object for each element, and sets each XML object's properties accordingly.

From that point on, the menu is in memory, and onCreateOptionsMenu() doesn't get called again until the activity is destroyed. Each time the user selects a menu item, though, onOptionsItemSelected() gets called to process that click. We'll talk about this in the next section.

You need to update the StatusActivity to load the options menu. To do that, add an onCreateOptionsMenu() method to StatusActivity. This method gets called only the first time the user clicks Menu, as Example 9-7 illustrates.

Example 9-7. onCreateOptionsMenu() callback of Main Activity

```
// Called to lazily initialize the action bar
@Override
public boolean onCreateOptionsMenu(Menu menu) { // ❶
    // Inflate the menu items to the action bar.
    getMenuInflater().inflate(R.menu.main, menu); // ❷
    return true; // ❸
}
```

❶ Called the first time this menu needs to be displayed, such as the first time this activity is rendered on the screen.

❷ Get the MenuInflater object from the context view getMenuInflater(). Then use the inflater to inflate the menu from the XML resource.

❸ You must return true for this menu to be displayed.

Updating StatusActivity to Handle Menu Events

You also need a way to handle various clicks on the menu items. To do that, add another callback method, onOptionsItemSelected(). This method is called every time the user clicks a menu item, as Example 9-8 illustrates.

Example 9-8. Main Activity, final

```
// Called when an options item is clicked
package com.marakana.android.yamba;

import android.app.Activity;
import android.content.Intent;
import android.os.Bundle;
import android.view.Menu;
import android.view.MenuItem;
import android.widget.Toast;

public class MainActivity extends Activity {

    @Override
    protected void onCreate(Bundle savedInstanceState) {
            super.onCreate(savedInstanceState);
            setContentView(R.layout.activity_main);
    }

    // Called to lazily initialize the action bar
    @Override
    public boolean onCreateOptionsMenu(Menu menu) {
            // Inflate the menu items to the action bar.
            getMenuInflater().inflate(R.menu.main, menu);
            return true;
    }

    // Called every time user clicks on an action
    @Override
    public boolean onOptionsItemSelected(MenuItem item) { // ❶
            switch (item.getItemId()) { // ❷
            case R.id.action_settings:
                    startActivity(new Intent(this, SettingsActivity.class)); // ❸
                    return true; // ❹
            case R.id.action_tweet:
                    startActivity(new Intent("com.marakana.android.yamba.action
                            .tweet"));
                    return true;
            default:
                    return false;
            }
    }
}
```

❶   onOptionsItemSelected() is called when user selects an item in the menu.

---

❷ Because the same method is called regardless of which item the user clicks, you need to figure out the ID of that item, and based on that, switch to a specific case to handle each item. At this point, we have only one menu item, but that might change in the future. Switching an item ID is a very scalable approach and will adapt nicely as our application grows in complexity.

❸ The startActivity() method in context launches a new activity. In this case, we are creating a new intent that specifies starting the SettingsActivity class. As you remember from "Intents" on page 68, intents are Android's way of specifying what is the target of your request to startActivity(). In this case, we're starting our settings activity.

❹ Return true to consume the event here.

 Just as before, you could use the Eclipse shortcut Source→Override/ Implement Methods to add both onCreateOptionsMenu() and onOp tionsItemSelected().

You should be able to run your application at this point and see the new SettingsAc tivity by clicking Menu → Settings in StatusActivity (see Figure 9-4). Try changing your username and password, then reboot your phone, restart the app, and verify that the information is still there.

Shared Preferences and Updating Status Fragment

Now that we have a preference activity and a way to save our username and password, it is time to make use of the preferences. To do so programmatically, use the SharedPre ference class provided by the Android framework.

This class is called SharedPreference because this preference is easily accessible from any component of this application (activities, services, broadcast receivers, and content providers).

In StatusFragment, add a definition for the prefs object globally to the class:

```
SharedPreferences prefs;
```

Now, to get the preference object, add the code in Example 9-9 to doInBackground().

Figure 9-4. SettingsActivity

Example 9-9. StatusFragment, reading username/password from the settings

```
...
    @Override
    protected String doInBackground(String... params) {
        try {
                SharedPreferences prefs = PreferenceManager
                    .getDefaultSharedPreferences(getActivity()); // ❶
                String username = prefs.getString("username", ""); // ❷
                String password = prefs.getString("password", "");

                // Check that username and password are not empty.
                // If empty, Toast a message to set login info and bounce
                // to SettingActivity.
                // Hint: TextUtils.
                if (TextUtils.isEmpty(username) ||
        TextUtils.isEmpty(password)) { // ❸
                    getActivity().startActivity(
                    new Intent(getActivity(), SettingsActivity.class));
                        return "Please update your username and password";
                }
```

```
                          YambaClient cloud = new YambaClient(username, password);
     // ❹
                          cloud.postStatus(params[0]);
  . . .
```

❶ Each application has its own shared preferences available to all components of this application context. To get the instance of this `SharedPreferences`, we use `PreferenceManager.getDefaultSharedPreferences()` and pass it this as the current context for this app. The name "shared" could be confusing. To clarify, it means that this preference object contains data shared by various parts of this application only; it is not shared with any other application.

❷ Get the username and password from the shared preference object. The first parameter in `getString()` is the key we assigned to each preference item, such as `username` and `password`. The second argument is the default value in case such a preference is not found. Keep in mind that the first time a user runs your application, the preference file doesn't exist, so defaults will be used. So, if the user hasn't set up her preferences in `SettingsActivity`, this code will attempt to log in with an empty username and password, and thus fail. However, the failure will happen when the user tries to do the actual status update because that's how the *yambaclientlib* library is designed.

❸ This is just a quick check that we actually have some legit values already configured in settings. If we don't, we'll communicate that to the user so she can go ahead and update her username and password first.

❹ Log in to the Yamba service with user-defined preferences.

 The default username and password for the yamba.marakana.com service is student/password. Again, we assume you are using the service at *http://yamba.marakana.com/* with the API root of *http://yamba.marakana.com/api* using login/password: **"student"/"password"**. Shhh, please don't tell anyone!

At this point, your app is working with user-specified login credentials. Next, we'll look at how the filesystem is organized on a typical Android device.

The Filesystem Explained

So, where does the device store these preferences? How secure are the username and password? To answer that, we need to look at how the Android filesystem is organized.

Exploring the Filesystem

There are two ways for you to access the filesystem on an Android device: via Eclipse or via the command line.

In Eclipse, use the File Explorer view to access the filesystem. To open up the File Explorer view, go to Window → Show View → Other → Android → File Explorer. You can also access the File Explorer view via the DDMS perspective (*http://develop er.android.com/guide/developing/tools/ddms.html*). Select the DDMS perspective icon in the top-right corner of Eclipse or go to Window → Open Perspective → Other → DDMS. If you have multiple devices connected to your workstation, make sure you select which one you are working with in the Devices view. You should now be able to navigate through the device's filesystem.

If you prefer the command line, you can always use `adb shell` to get to the shell of the device. From there you can explore the filesystem like you would on any other Unix platform. We'll show this use of the shell momentarily.

Filesystem Partitions

There are three main parts of the filesystem on every Android device. As shown in Figure 9-5, they are:

- The system partition (*/system/*)
- The SDCard partition (*/sdcard/*)
- The user data partition at (*/data/*)

System Partition

Your entire Android operating system is located in the system partition. This is the main partition that contains all your preinstalled applications, system libraries, Android framework, Linux command-line tools, and so on.

The system partition is mounted read-only, meaning that you as developer have very little influence over it. As such, this partition is of limited interest to us.

The system partition in the Emulator corresponds to the *system.img* file in your platform images directory, located in the *android-sdk/platforms/android-8/images* folder.

SDCard Partition

The SDCard partition is a free-for-all mass storage area. Your app can read files from this partition as well as write files to it if it holds the `WRITE_TO_EXTERNAL_STORAGE`

---

permission. This is a great place to store large files, such as music, photos, videos, and similar items.

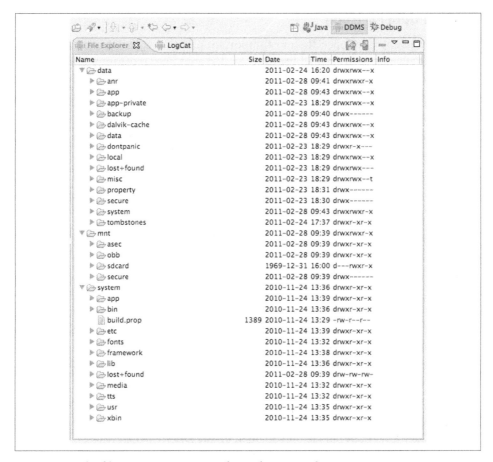

Name	Size	Date	Time	Permissions	Info
▼ 🗀 data		2011-02-24	16:20	drwxrwx--x	
▶ 🗀 anr		2011-02-28	09:41	drwxrwx-x	
▶ 🗀 app		2011-02-28	09:43	drwxrwx--x	
▶ 🗀 app-private		2011-02-23	18:29	drwxrwx--x	
▶ 🗀 backup		2011-02-28	09:40	drwx------	
▶ 🗀 dalvik-cache		2011-02-28	09:43	drwxrwx--x	
▶ 🗀 data		2011-02-28	09:43	drwxrwx--x	
▶ 🗀 dontpanic		2011-02-23	18:29	drwxr-x---	
▶ 🗀 local		2011-02-23	18:29	drwxrwx--x	
▶ 🗀 lost+found		2011-02-23	18:29	drwxrwx---	
▶ 🗀 misc		2011-02-23	18:29	drwxrwx--t	
▶ 🗀 property		2011-02-23	18:31	drwx------	
▶ 🗀 secure		2011-02-23	18:30	drwx------	
▶ 🗀 system		2011-02-28	09:43	drwxrwxr-x	
▶ 🗀 tombstones		2011-02-24	17:37	drwxr-xr-x	
▼ 🗀 mnt		2011-02-28	09:39	drwxrwxr-x	
▶ 🗀 asec		2011-02-28	09:39	drwxr-xr-x	
▶ 🗀 obb		2011-02-28	09:39	drwxr-xr-x	
▶ 🗀 sdcard		1969-12-31	16:00	d---rwxr-x	
▶ 🗀 secure		2011-02-28	09:39	drwx------	
▼ 🗀 system		2010-11-24	13:36	drwxr-xr-x	
▶ 🗀 app		2010-11-24	13:39	drwxr-xr-x	
▶ 🗀 bin		2010-11-24	13:36	drwxr-xr-x	
📄 build.prop	1389	2010-11-24	13:29	-rw-r--r--	
▶ 🗀 etc		2010-11-24	13:39	drwxr-xr-x	
▶ 🗀 fonts		2010-11-24	13:32	drwxr-xr-x	
▶ 🗀 framework		2010-11-24	13:38	drwxr-xr-x	
▶ 🗀 lib		2010-11-24	13:36	drwxr-xr-x	
▶ 🗀 lost+found		2011-02-28	09:39	drw-rw-rw-	
▶ 🗀 media		2010-11-24	13:32	drwxr-xr-x	
▶ 🗀 tts		2010-11-24	13:32	drwxr-xr-x	
▶ 🗀 usr		2010-11-24	13:35	drwxr-xr-x	
▶ 🗀 xbin		2010-11-24	13:35	drwxr-xr-x	

Figure 9-5. The filesystem as seen via File Explorer in Eclipse

 Starting with the FroYo version of Android, the */sdcard* mount point appears in the Eclipse File Explorer under the */mnt/sdcard* location. This is due to the new feature in FroYo that allows for storing and running applications on the SDCard as well.

As an app developer, the SDCard partition is very useful and important to you. At the same time, this partition is not very structured.

This partition typically corresponds to *sdcard.img* in your Android Virtual Device (AVD) directory. This directory is in your *~/.android/avd/* folder and will have a subdirectory for each specific virtual device. On the physical device, it is an actual SD card (*http://en.wikipedia.org/wiki/Secure_Digital*).

The User Data Partition

As a user and app developer, the most important partition is the user data partition. This stores all your user data, all the downloaded apps, and most importantly, all the applications' data. This includes preinstalled apps as well as user-downloaded apps.

So, while user apps are stored in the */data/app* folder, the most important folder to us as app developers is the */data/data* folder. Within this folder is a subfolder corresponding to each app. This folder is identified by the Java package that this app used to sign itself. Again, this is why Java packages are important to Android security.

The Android framework provides a number of handy methods as part of its context that help you access the user data filesystem from within your application. For example, take a look at `getFilesDir()` (*http://bit.ly/1gc9JKq*).

The user data partition typically corresponds to *user-data.img* in your Android Virtual Device (AVD) directory. As before, this directory is in your *~/.android/avd/* folder and will have a subdirectory for each specific virtual device.

When you create a new app, you assign your Java code to a specific package. Typically, this package follows the Java convention of reverse domain name plus app name. For example, the Yamba app is in the `com.marakana.android.yamba` package. So, once installed, Android creates a special folder just for this app under */data/data/com.marakana.android.yamba/*. This folder is the cornerstone of the private, secured filesystem dedicated to each app.

There will be subfolders in */data/data/com.marakana.android.yamba/*, but they are well defined. For example, the preferences are in */data/data/com.marakana.android.yamba/shared_prefs/*. As a matter of fact, if you open up the DDMS perspective in Eclipse and select File Explorer, you can navigate to this folder. You will probably see the *com.marakana.android.yamba_preferences.xml* file in there. You could pull this file and examine it, or use `adb shell`.

`adb shell` is common `adb` subcommands to access the shell of your device (either physical or virtual). For instance, you could just open up your command-line terminal and type:

```
[user:~]> adb shell
# cd /data/data/com.marakana.android.yamba/shared_prefs
# cat com.marakana.android.yamba_preferences.xml
<?xml version='1.0' encoding='utf-8' standalone='yes' ?>
<map>
<string name="password">password</string>
<string name="username">student</string>
```

```
</map>
#
```

This XML file represents the storage for all our preference data for this application. As you can see, our username, password, and API root are all stored in there.

Filesystem Security

So, how secure is this? This is a common question posed by security folks. Storing usernames and passwords in clear text always raises eyebrows.

To answer this question, we usually compare it to finding someone's laptop on the street. Although we can easily gain access to the "hard drive" via the adb tool, that doesn't mean we can read its data. Each folder under */data/data* belongs to a separate user account managed by Linux. Unless our app is that app, it won't have access to that folder. So, short of an intruder reading byte by byte on the physical device, even clear-text data is secure.

On the Emulator, we have root permissions, meaning we can explore the entire filesystem. This is useful for development purposes.

Summary

At this point, the user can specify her username and password for the micro-blogging site. This makes the app usable to way more people than the previous version in which this information was hardcoded.

Figure 9-6 illustrates what we have done so far as part of the design outlined earlier in Figure 6-4.

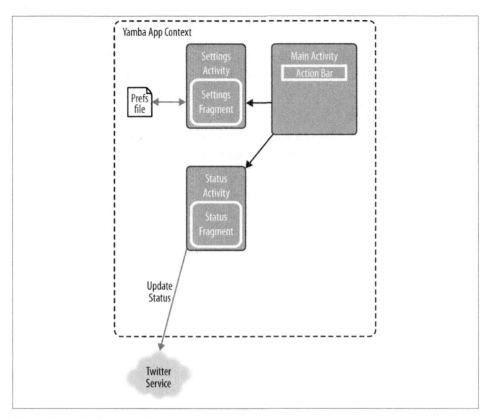

Figure 9-6. Yamba completion

Services

Services are among the main building blocks in Android. Unlike an activity, a service doesn't have a user interface; it is simply a piece of code that runs in the background of your application.

Services are used for processes that should run independently of activities, which may come and go. Our Yamba application, for example, needs to create a service to periodically connect to the cloud and check for new statuses from the user's friends. This service will be always on and always running, regardless of whether the user ever starts the activity.

Just like an activity, a service has a well-defined life cycle. As the developer, you get to define what happens during transitions between states. Whereas an activity's state is managed by the runtime's `ActivityManager`, service state is controlled more by intents. First, you must create the service. Whenever an activity needs the service, the activity will invoke the service through an intent, as described in "Intents" on page 68. This is called *starting* the service. A running service can receive the start message repeatedly and at unanticipated times. You can also stop a service, which is called *destroying* it.

A service can be bound or unbound. Bound services can provide more specific APIs to other applications via an interface called the Android Interface Definition Language (AIDL). We'll focus on unbound services, where the life cycle of a service is not tied to the life cycle of the activities that started them. The only states for unbound services are started and stopped (destroyed).

In this chapter, you will create a service. The purpose of this service is to run in the background and update your app with the latest timeline from the user's Yamba account. Initially, the service will just print your friends' timeline to the logfile. The service will create a separate thread, so you will learn about concurrency in this chapter as well. You will also learn about toasts and understand the context in which services and activities run.

By the end of this chapter, you will have a working app that can both post to Yamba and periodically check what friends are up to.

Our Example Service: RefreshService

As mentioned in the introduction to this chapter, we need a service to run as an always-on background process, pulling the latest Yamba statuses into a local database. The purpose of this pull mechanism is to cache updates locally so our app can have data even when it's offline. We'll call this service `RefreshService`.

Steps to creating a service are:

1. Create the Java class representing your service.
2. Register the service in the *AndroidManifest.xml* file.
3. Start the service.

Creating the RefreshService Java Class

The basic procedure for creating a service, as with activities and other main building blocks, is to subclass a `Service` class provided by the Android framework.

To create the new service, we need to create a new Java file. Go ahead and select your Java package in the *src* folder, right-click and choose New→Class, and type in **Refresh Service** as the class name. This will create a new *RefreshService.java* file as part of your package.

You may recall from "Services" on page 68 that a typical unbound service goes through the life cycle illustrated in Figure 10-1.

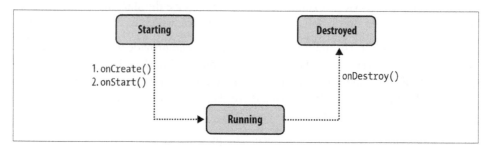

Figure 10-1. Service life cycle

Next, we want to override some of the main life cycle methods:

onCreate()
>   Called when the service is created for the first time

```
onStartCommand()
```
   Called each time the service is started

```
onDestroy()
```
   Called when the service is terminated

To do that, you can use the Eclipse tool Source→Override/Implement Methods and select those three methods.

At this point, in the spirit of producing a minimally working app at each stage of learning, we'll write just a little code that logs a note in each of the overridden methods. So the shell of our service looks like the code in Example 10-1.

Example 10-1. RefreshService.java, version 1

```java
package com.marakana.android.yamba;

import android.app.Service;
import android.content.Intent;
import android.os.IBinder;
import android.util.Log;

public class RefreshService extends Service {
  static final String TAG = "RefreshService"; // ❶

  @Override
  public IBinder onBind(Intent intent) { // ❷
    return null;
  }

  @Override
  public void onCreate() { // ❸
    super.onCreate();
    Log.d(TAG, "onCreated");
  }

  @Override
  public int onStartCommand(Intent intent, int flags, int startId) { // ❹
    super.onStartCommand(intent, flags, startId);
    Log.d(TAG, "onStarted");
    return START_STICKY;
  }

  @Override
  public void onDestroy() { // ❺
    super.onDestroy();
    Log.d(TAG, "onDestroyed");
  }
}
```

❶ As in all major classes, we like to add the TAG constant because we use Log.d()
quite a bit.

❷ onBind() is used in bound services to return the actual implementation of
something called a *binder*. Because we are not using a bound service, we can just
return null here.

❸ onCreate() is called when the service is initially created. It is not called for
subsequent startService() calls, so it is a good place to do work that needs to
be done only once during the life of a service.

❹ onStartCommand() is called each time the service receives a startService()
intent. A service that is already started could get multiple requests to start again,
and each will cause onStartCommand() to execute. START_STICKY is used as a
flag to indicate this service is started and stopped explicitly, which is what we
want in our case.

❺ onDestroy() is called just before the service is destroyed by the stopSer
vice() request. This is a good place to clean up things that might have been
initialized in onCreate().

Introducing IntentService

It's important to note that your service is going to run on the main thread of the application, i.e., the UI thread. Because our service is going to be connecting to the cloud to
pull down the latest data, we once again have the problem of networking on the UI
thread.

One solution to this problem would be to handle our own Thread or Runnable, but this
gets messy. It turns out that Android SDK provides an alternative subclass of service,
called *IntentService*.

An intent service is similar to regular service, with two main exceptions: whatever work
is to be done in onHandleIntent() will execute on a separate worker thread, and once
it's done, the service will stop.

Example 10-2 shows a minimal version of an intent service.

Example 10-2. RefreshService.java using IntentService

```
package com.marakana.android.yamba;

import android.app.IntentService;
import android.content.Intent;
import android.os.IBinder;
import android.util.Log;

public class RefreshService extends IntentService {
    static final String TAG = "RefreshService"; // ❶
```

```
    public RefreshService() { // ❷
            super(TAG);
        }


    @Override
    public void onCreate() { // ❸
        super.onCreate();
        Log.d(TAG, "onCreated");
    }

    // Executes on a worker thread
        @Override
        protected void onHandleIntent(Intent intent) { // ❹
        Log.d(TAG, "onStarted");
    }

    @Override
    public void onDestroy() { // ❺
        super.onDestroy();
        Log.d(TAG, "onDestroyed");
    }
}
```

❶   This is the usual tag that we'll use for logging.

❷   IntentService requires a default constructor. In that constructor, you need to call super() and pass a name of this service. TAG variable comes in handy for this.

❸   Just as in a regular service, onCreate() is called when the service is created for the first time.

❹   onHandleIntent() is where we do the main work. This work will be executed on a separate thread. This is one of the main differences between a service and an intent service.

❺   Just as in a regular service, onDestroy() is called when the service is about to be stopped. Unlike a regular service, onDestroy() is called as soon as onHand leIntent() terminates.

Update the Manifest File

Now that we have the shell of our service, we have to define it in the manifest file, just like any other main building block; otherwise, we won't be able to call our service. Simply open *AndroidManifest.xml*, click the rightmost tab to see the raw XML code, and add the following <service> tag within the <application> element:

```
...
    <application android:icon="@drawable/icon" android:label="@string/app_name">
        ...
        <service android:name=".RefreshService" /> <!-- ❶ -->
        ...
    </application>
...
```

❶   `RefreshService` definition.

Services are equal in importance to activities as Android building blocks, so they appear at the same level in the manifest file.

Add Menu Items

Now that we have defined and declared the service, we need a way to start and stop it. The easiest way would be to add a menu button to our options menu that we have already created. Later on, we'll have a more intelligent way of starting services, but for now this manual approach is easier to understand.

To add start/stop menu buttons, we'll add two more menu items to our *menu.xml* resource, just as we created the `Prefs` menu item in "The Menu Resource" on page 150. The updated *menu.xml* now looks like Example 10-3.

Example 10-3. menu.xml

```
<?xml version="1.0" encoding="utf-8"?>
<menu xmlns:android="http://schemas.android.com/apk/res/android">
  <item android:id="@+id/itemPrefs" android:title="@string/titlePrefs"
    android:icon="@android:drawable/ic_menu_preferences"></item>  <!-- ❶ -->
  <item android:title="@string/titleRefresh"
    android:id="@+id/itemServiceStart"
    android:icon="@android:drawable/ic_menu_rotate"></item>  <!-- ❷ -->
</menu>
```

❶   This is the item we defined in the previous chapter.

❷   The `ServiceStart` item has the usual `id`, `title`, and `icon` attributes. This icon is another Android system resource.

Now that the menu resource has been updated, it's time to handle those items when the user clicks them.

Update the Options Menu Handling

To handle new menu items, we need to update the `onOptionsItemSelected()` method in `StatusActivity`, just as we did in "Updating StatusActivity to Handle Menu Events" on page 154. So open your *StatusActivity.java* file and locate the `onOptionsI` temSelected() method. We now have a framework in this method to support any number of menu items. To add support for starting and stopping our service, we launch intents pointing to our RefreshService via the startService() call. The final code looks like this:

```
// Called when an options item is clicked
@Override
public boolean onOptionsItemSelected(MenuItem item) {
  switch (item.getItemId()) {
  case R.id.itemRefresh:
    startService(new Intent(this, RefreshService.class)); // ❶
    break;
  case R.id.itemPrefs:
    startActivity(new Intent(this, PrefsActivity.class));
    break;
  default:
    return false;
  }

  return true;
}
```

❶ Creates an intent to start RefreshService. If the service doesn't already exist, the runtime calls the service's onCreate() method. Then onStartCommand() is called, regardless of whether this service is new or already running.

In this example, we are using explicit intents (explained in "Intents" on page 68) to specify exactly which class the intents are intended for, namely *RefreshService.class*.

Testing the Service

At this point, you can restart your application. Note that you do not need to restart the emulator. When your application starts up, click the menu, and your new buttons should appear in the menu options. You can now freely click the start and stop service buttons.

To verify that your service is working, open up your LogCat and look for the appropriate log messages that you generated in your service code. Remember from "Logging Messages in Android" on page 108 that you can view the LogCat both in Eclipse and via the command line.

Your service is now working, although it's not doing much at this point.

Pulling Data from Yamba

We now have a framework and are ready to make the actual connection to the online Twitter-like service, pull the status data, and display that data in our application. Twitter and Twitter-like services offer many different APIs to retrieve our friends' updates. The *yambaclientlib.jar* library exposes most of them to us via the `YambaClient` class. Perhaps one of the most appropriate methods is `getTimeline()`, which returns the 20 most recent posts made over the past 24 hours from the user and her friends.

To use this Twitter API feature, we need to connect to the online service. And to do that, we need the username, password, and root API for our online service. We've written most of this code before when we needed to post to Twitter API as well.

Now we can write new code for `RefreshService` and have it connect to the online API to pull the latest status updates from our friends. Example 10-4 shows the final version.

Example 10-4. RefreshService.java, final version

```
package com.marakana.android.yamba;
package com.marakana.android.yamba;

import java.util.List;

import android.app.IntentService;
import android.content.ContentValues;
import android.content.Intent;
import android.content.SharedPreferences;
import android.net.Uri;
import android.preference.PreferenceManager;
import android.text.TextUtils;
import android.util.Log;
import android.widget.Toast;

import com.marakana.android.yamba.clientlib.YambaClient;
import com.marakana.android.yamba.clientlib.YambaClient.Status;
import com.marakana.android.yamba.clientlib.YambaClientException;

public class RefreshService extends IntentService {
    private static final String TAG = "RefreshService";

        public RefreshService() {
                super(TAG);
        }

        @Override
        public void onCreate() {
                super.onCreate();
                Log.d(TAG, "onCreated");
        }

        // Executes on a worker thread
        @Override
        protected void onHandleIntent(Intent intent) {
                SharedPreferences prefs = PreferenceManager
                                .getDefaultSharedPreferences(this);      // ❶
                final String username = prefs.getString("username", "");
                final String password = prefs.getString("password", "");
```

```
        // Check that username and password are not empty
        if (TextUtils.isEmpty(username) || TextUtils.isEmpty(password)) {
// ❷
                Toast.makeText(this,
                "Please update your username and password",
                        Toast.LENGTH_LONG).show();
                return;
        }
        Log.d(TAG, "onStarted");

        YambaClient cloud = new YambaClient(username, password);    // ❸
        try {
                List<Status> timeline = cloud.getTimeline(20);  // ❹
                for (Status status : timeline) {     // ❺
                        Log.d(TAG,
                                String.format("%s: %s", status.getUser(),
                                        status.getMessage())));  // ❻
                }

        } catch (YambaClientException e) {  // ❼
                Log.e(TAG, "Failed to fetch the timeline", e);
                e.printStackTrace();
        }

        return;
    }

    @Override
    public void onDestroy() {
        super.onDestroy();
        Log.d(TAG, "onDestroyed");
    }
}
```

❶ We get the reference to the shared preferences and extract the username and password. This is the same code as we did when posting to the cloud in "Shared Preferences and Updating Status Fragment" on page 155.

❷ This is a minimal check to make sure username and password are not empty. If they are, we tell the user via a toast, and return, meaning the service stops right here.

❸ We get the reference to YambaClient object so we can connect to the cloud.

❹ We call getYambaClient() in YambaApplication to get the yamba object, and then call getTimeline() on it to get the last 20 status posts from the past 24 hours. Note that this is the actual method that implements the web service call to our cloud service. As such, it could take some time to complete, depending on the network latency. Because we run this in our dedicated thread, we won't affect the main user interface thread while we wait for the network operation to complete. We are using Java generics (*http://en.wikipedia.org/wiki/Generics_in_Java*) to define the timeline variable as a List of Status instances.

❺ Now that we have initialized the timeline list, we can loop over it. The easiest approach is to use Java's "for each" loop, which automatically iterates over our list, assigning each element in turn to the status variable.

❻ For now, we simply print out the statuses of who said what to the LogCat output.

❼ A network call can fail for any number of reasons. Here we handle failure by printing the stack trace of what went wrong. The actual printout will be visible in LogCat.

Testing the Service

Now we can run our application, click Refresh in the menu to start the service, and see the list of our friends' statuses in the LogCat:

```
D/RefreshService(  310): Marko Gargenta: it is great that you got my message
D/RefreshService(  310): Marko Gargenta: hello this is a test message from my phone
D/RefreshService(  310): Marko Gargenta: Test
D/RefreshService(  310): Marko Gargenta: right!
...
```

Summary

We now have a working service, which we start manually whenever we want to refresh the data. The service connects to the cloud service and pulls down the status posts from our friends. For now, we just print this data in the LogCat, but in the next chapter we'll insert the data into the database.

Figure 10-2 illustrates what we have done so far as part of the design outlined earlier in Figure 6-4.

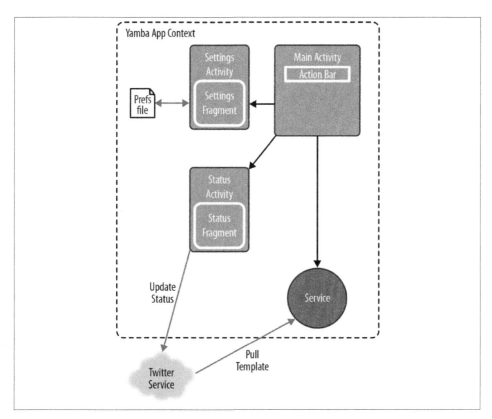

Figure 10-2. Yamba completion

Content Providers

In this chapter, you will learn about content providers, one of the main building blocks of Android. In a nutshell, think of a content provider as an interface to your app's data. Before we get to that, we need the data. So, we'll talk about databases first.

In this chapter, you will learn how Android supports databases. You will learn to create and use a database inside the Yamba application to store your status updates locally. Local data will help Yamba display statuses to the user quickly, without having to wait for the network to provide the data. Our service will run in the background and periodically update the database so that the data is relatively fresh. This will improve the overall user experience of the application.

Databases on Android

The Android system uses databases to store useful information that needs to be persisted even when the user kills the app or even shuts down the device and powers it back on. The data includes contacts, system settings, bookmarks, and so on.

So, why use a database in a mobile application? After all, isn't it better to keep our data in a cloud where it's always backed up instead of storing it in a mobile device that is easily lost or damaged?

A database in a mobile device is very useful as a supplement to the online world. Although in many cases it is much better to count on the data living in the cloud, it is useful to store it locally in order to access it more quickly and have it available even when the network is not available. In this case, we are using a local database as a cache. This is also how we use it in our Yamba application.

About SQLite

SQLite is an open source database that has been around for a long time, is quite stable, and is popular on many small devices, including Android. There are couple of good reasons why SQLite is a great fit for Android app development:

- It's a zero-configuration database. That means there's absolutely no database configuration for you as the developer. This makes it relatively simple to use.

- It doesn't have a server. There's no SQLite database process running. It is basically a set of libraries that provides the database functionality. Not having a server to worry about is also a good thing.

- It's a single-file database. This makes database security straightforward, because it boils down to filesystem security. We already know that Android sets aside a special, secure sandbox for each application.

- It's open source.

The Android framework offers several ways to use SQLite easily and effectively, and we'll look at the basic usage in this chapter. You may be pleased to find that, although SQLite uses SQL, Android provides a higher-level library with an interface that is much easier to integrate into an application.

 Although SQLite support is built into Android, it is by no means your only option when it comes to data persistence for your app. You can always use another database system, such as JavaDB or MongoDB, but you'd have to bundle the required libraries with your app and would not be able to rely on Android's built-in database support. SQLite is not an alternative to a full SQL server; instead, it is an alternative to using a local file with an arbitrary format.

DbHelper

Android provides an elegant interface for your app to interact with an SQLite database. To access the database, you first need a helper class that provides a "connection" to the database, creating the connection if it doesn't already exist. This class, provided to you by the Android framework, is called SQLiteOpenHelper. The database class it returns is an instance of SQLiteDatabase.

The following subsections explain some of the background concepts you should understand when working on building DbHelper, a class that extends SQLiteOpenHelper. We won't explain SQL or basic database concepts such as normalization, because there are hundreds of good places to find that information, and we expect most of our readers already know it. However, this chapter should give you enough to get started, even if your knowledge of databases is spotty.

The Database Schema and Its Creation

A schema is just a description of what's in a database. In our Yamba database, for instance, we want fields for the following information about each message we retrieve from Yamba:

user
> The user who sent the message

message
> The text of the message

created_at
> The date when the message was sent

So each row in our table will contain the data for one message, and these three items will be the columns in our schema, along with a unique ID for each tweet. We need the ID so we can easily refer to a tweet. SQLite, like most databases, allows us to declare the ID as a primary key and even assigns a unique number automatically to each tweet for us.

The schema has to be created when our application starts, so we'll do it in the on Create() method of DbHelper. We might add new fields or change existing ones in a later version of our application, so we'll assign a version number to our schema and provide an onUpgrade() method that we can call to alter the schema.

onCreate() and onUpgrade() are the only methods in our application when we need to use SQL. We'll execute CREATE TABLE in onCreate() to create a table in our database. In a production application, we'd use ALTER TABLE in onUpgrade() when the schema changes, but that requires a lot of complex introspection of the database, so for now we'll use DROP TABLE and recreate the table. Of course, DROP TABLE destroys any data currently in the table, but that's not a problem for our Yamba application. It always refills the table with tweets from the past 24 hours, which are the only ones our users will care about.

Four Major Operations

The DbHelper class offers you a high-level interface that's much simpler than SQL. The developers realized that most applications use databases for only four major operations, which go by the appealing acronym CRUD: create, read (query), update, and delete. To fulfill these requirements, DbHelper offers:

insert()
> Inserts one or more rows into the database

query()
> Requests rows matching the criteria you specify

`update()`
Replaces ones or more rows that match the criteria you specify

`delete()`
Deletes rows matching the criteria you specify

Each of these methods has variants that enhance it with other functions. To use one of the methods, create a `ContentValues` container and place in it the information you want inserted, updated, etc. This chapter will show you the process for an insert, and the other operations work in similar ways.

So, why not use SQL directly? There are several good reasons:

- From a security point of view, a SQL statement is a prime candidate for a security attack on your application and data, known as a SQL injection attack (*http://en.wiki pedia.org/wiki/SQL_injection*). That is because the SQL statement takes user input, and unless you check and isolate it very carefully, this input could embed other SQL statements with undesirable effects.

- From a performance point of view, executing SQL statements repeatedly is highly inefficient because you'd have to parse the SQL every time the statement runs.

- The `DbHelper` methods are more robust and less likely to pass through the compiler with undetected errors. When you include SQL in a program, it's easy to create errors that turn up only at runtime.

Android's database framework supports only prepared statements for standard CRUD operations: `INSERT`, `UPDATE`, `DELETE`, and `SELECT`. If you need to execute other SQL statements, you need to pass them directly to SQLite. That's why we used `execSQL()` to run the code to `CREATE TABLE`. This is OK because that code doesn't depend on any user input, and as such SQL injection is not possible. Additionally, that code runs very rarely, so there's no need to worry about the performance implications.

Cursors

A query returns a set of rows along with a pointer called a *cursor*. You can retrieve results one at a time from the cursor, causing it to advance each time to the next row. You can also move the cursor around in the result set. An empty cursor indicates that you've retrieved all the rows.

In general, anything you do with SQL could lead to an SQL exception because its code is interacting with a system that's outside of our direct control. For example, the database could be running out of space or somehow corrupted. So, it is a good practice to handle all the `SQLExceptions` by surrounding your database calls in try/catch blocks.

It's easy to do this using the Eclipse shortcut:

1. Select the code for which you'd like to handle exceptions. Typically this would be most of your SQL calls.

2. In the Eclipse menu, choose Source→Surround With→Try/catch Block. Eclipse will generate the appropriate try/catch statements around your code for the proper exception class.

3. Handle this exception in the `catch` block. This might be a simple call to `Log.e()` to pass the tag, message, and the exception object itself.

Status Contract Class

It turns out that we'll need to have a whole bunch of constants that identify things like the name and version of our database, the column names, the table the data lives in, and so on. It is the best practice to store these constants in a separate class, usually named something-Contract.

Example 11-1 shows the `StatusContract` at this point in our development.

Example 11-1. StatusContract.java

```java
package com.marakana.android.yamba;

import android.net.Uri;
import android.provider.BaseColumns;

public class StatusContract {

    // DB specific constants
        public static final String DB_NAME = "timeline.db"; // ❶
        public static final int DB_VERSION = 1; // ❷
        public static final String TABLE = "status"; // ❸

        public static final String DEFAULT_SORT = Column.CREATED_AT + " DESC"; // ❹

        public class Column { // ❺
                public static final String ID = BaseColumns._ID; // ❻
                public static final String USER = "user";
                public static final String MESSAGE = "message";
                public static final String CREATED_AT = "created_at";
        }
}
```

❶ This is the actual SQLite file that will contain the database.

❷ Database schemas are versioned. You can give it any version number, so we'll start with 1.

❸ This is the actual SQL table that will contain the data.

❹   The default sort order will sort by the timestamp (the CREATED_AT column), with latest status showing up first.

❺   These will be our column names.

❻   Although the ID could be any name, there's a convention in Android to name it _id, for which it provides an API-level contract as well. You should try to use this whenever you define an ID field.

So now we're going to create our own helper class to help us open our Yamba database (see Example 11-2). We'll call the class DbHelper. It will create the database file if one doesn't already exist, or it will upgrade the user's database if the schema has changed between versions.

Like many other classes in Android, we usually start by subclassing a framework class, in this case SQLiteOpenHelper. We then need to implement the class's constructor, as well as onCreate() and onUpgrade() methods.

Example 11-2. DbHelper.java, version 1

```
package com.marakana.android.yamba;

import android.content.Context;
import android.database.sqlite.SQLiteDatabase;
import android.database.sqlite.SQLiteOpenHelper;
import android.util.Log;

public class DbHelper extends SQLiteOpenHelper { // ❶
    private static final String TAG = DbHelper.class.getSimpleName();

        public DbHelper(Context context) {
                super(context, StatusContract.DB_NAME, null, StatusContract.DB_VERSION);
        // ❷
        }

        // Called only once first time we create the database
        @Override
        public void onCreate(SQLiteDatabase db) {
                String sql = String
                                .format("create table %s (%s int primary key, %s text,
                        %s text, %s int)",
                                                StatusContract.TABLE,
                        StatusContract.Column.ID,
                                                StatusContract.Column.USER,
                                                StatusContract.Column.MESSAGE,
                                                StatusContract.Column.CREATED_AT);
                        // ❸
                Log.d(TAG, "onCreate with SQL: "+sql);
                db.execSQL(sql); // ❹
        }
```

```
            // Gets called whenever existing version != new version, i.e. schema changed
            @Override
            public void onUpgrade(SQLiteDatabase db, int oldVersion, int newVersion) {
// ❺
                    // Typically you do ALTER TABLE ...
                    db.execSQL("drop table if exists " + StatusContract.TABLE);
                    onCreate(db);
            }
}
```

❶   Start by subclassing SQLiteOpenHelper.

❷   We override the constructor and pass in the database name and version from
     our contracts class. Doing this explicitly in code is reasonable because this
     information doesn't change all that much.

❸   This is the actual SQL that we'll pass on to the database to have it create the
     appropriate SQL schema that we need. We plug five strings we defined earlier
     into an SQL statement.

❹   Once we have a string containing an SQL statement that creates the database,
     run execSQL() on the database object that was passed into onCreate().

❺   onUpgrade() is called whenever the user's database version is different from the
     application version. This typically happens when you change the schema and
     release the application update to users who already have older version of your
     app.

 As mentioned earlier, you would typically execute an SQL statement, ALTER TABLE, in onUpgrade(). Because we don't have an old
database to alter, we are assuming this application is still in prere-
lease mode and are just deleting any user data when recreating the
database.

Next, we need to update the service in order to have it open up the database connection,
fetch the data from the network, and insert it into the database.

Update RefreshService

Remember that our RefreshService connects to the cloud and gets the data. So Re
freshService also is responsible for inserting this data into the local database.

In Example 11-3, we update the RefreshService to pull the data from the cloud and
store it in the database.

Example 11-3. RefreshService.java, version 1

```java
package com.marakana.android.yamba;

import java.util.List;

import android.app.IntentService;
import android.content.ContentValues;
import android.content.Intent;
import android.content.SharedPreferences;
import android.net.Uri;
import android.preference.PreferenceManager;
import android.text.TextUtils;
import android.util.Log;
import android.widget.Toast;

import com.marakana.android.yamba.clientlib.YambaClient;
import com.marakana.android.yamba.clientlib.YambaClient.Status;
import com.marakana.android.yamba.clientlib.YambaClientException;

public class RefreshService extends IntentService {
    private static final String TAG = RefreshService.class.getSimpleName();

    public RefreshService() {
        super(TAG);
    }

    @Override
    public void onCreate() {
        super.onCreate();
        Log.d(TAG, "onCreated");
    }

    // Executes on a worker thread
    @Override
    protected void onHandleIntent(Intent intent) {
        SharedPreferences prefs = PreferenceManager
                        .getDefaultSharedPreferences(this);
        final String username = prefs.getString("username", "");
        final String password = prefs.getString("password", "");

        // Check that username and password are not empty
        if (TextUtils.isEmpty(username) || TextUtils.isEmpty(password)) {
            Toast.makeText(this,
                "Please update your username and password",
                            Toast.LENGTH_LONG).show();
            return;
        }
        Log.d(TAG, "onStarted");

    DbHelper dbHelper = new DbHelper(this); // ❶
    SQLiteDatabase db = dbHelper.getWritableDatabase(); // ❷
```

```
                ContentValues values = new ContentValues(); // ❸

                YambaClient cloud = new YambaClient(username, password);
                try {
                        List<Status> timeline = cloud.getTimeline(20);
                        for (Status status : timeline) {
                                values.clear(); // ❹
                                values.put(StatusContract.Column.ID, status.getId());
                                values.put(StatusContract.Column.USER,
                status.getUser());
                                values.put(StatusContract.Column.MESSAGE,
                status.getMessage());
                                values.put(StatusContract.Column.CREATED_AT, status
                                        .getCreatedAt().getTime());

                db.insertWithOnConflict(StatusContract.TABLE, null, values,
                        SQLiteDatabase.CONFLICT_IGNORE);// ❺

                        }

                } catch (YambaClientException e) {
                        Log.e(TAG, "Failed to fetch the timeline", e);
                        e.printStackTrace();
                }

                return;
        }

        @Override
        public void onDestroy() {
                super.onDestroy();
                Log.d(TAG, "onDestroyed");
        }
}
```

❶ Create the instance of DbHelper and pass this as its context. This works because the Android Service class is a subclass of Context. DbHelper will figure out whether the database needs to be created or upgraded.

❷ Get the writable database so we can insert new statuses into it. The first time we make this call, onCreate() in DbHelper will run and create the database file for this user.

❸ ContentValues is a simple data structure consisting of name-value pairs that map database table names to their respective values.

❹ For each record, we create a content value. We are reusing the same Java object, clearing it each time we start the loop and populating appropriate values for the status data.

❺ We insert the content value into the database via an `insertWithOnConflict()` call to the `SQLiteDatabase` object. Notice that we are not piecing together an SQL statement here, but rather using a *prepared statement* approach to inserting into the database.

A word about `insert()` versus `insertWithOnConflict()`. Because we keep asking for the latest timeline from the cloud, we'll be getting statuses that we've already inserted into the database. If we blindly tried to insert the old statuses, we'd violate the requirement that the ID for each status be unique, and the database would complain. Because we know this is going to be the case, we use `insertWithOnConflict()`, which allows us to specify what to do in case of a constract violation (e.g., duplicate ID). Our call passes the `CONFLICT_IGNORE` parameter to tell the database to just silently ignore our attempt to update it.

We are now ready to run our code and test it to make sure everything works.

Testing the Service

At this point, we can test whether the database was created properly and whether the service has populated it with some data. We're going to do this step by step.

Verify that the database was created

If the database file was created successfully, it will be located in the */data/data/com.marakana.android.yamba/databases/timeline.db* file. You can use the Eclipse DDMS perspective and File Explorer view to look at the filesystem of the device, or you can use `adb shell` on your command line, and then run this to make sure the file is there:

```
ls /data/data/com.marakana.android.yamba/databases/timeline.db
```

To use File Explorer in Eclipse, either open the DDMS perspective in the top-right corner of your Eclipse or go to Windows → Show View → Other → Android → File Explorer. This will open the view of the filesystem of the device you are currently looking at.

So far, you know that the database file is there, but don't really know whether the database schema was created properly. The next section addresses that.

Using sqlite3

Android ships with the command-line tool `sqlite3`. This tool gives you access to the database itself.

To see whether your database schema was created properly:

1. Open up your terminal or command-line window.

2. Type **adb shell** to connect to your running emulator or physical phone.

3. Change the directory to the location of your database file by typing **cd /data/data/com.marakana.android.yamba/databases/**.

4. Connect to the database with the sqlite3 timeline.db command.

At this point, you should be connected to the database. Your prompt should be sqlite>, indicating that you are inside the SQLite database:

```
[user:~]> adb shell
# cd /data/data/com.marakana.android.yamba/databases/
# ls
timeline.db
# sqlite3 timeline.db
SQLite version 3.6.22
Enter ".help" for instructions
Enter SQL statements terminated with a ";"
sqlite>
```

At this point, you can send two types of commands to your SQLite database:

- Standard SQL commands, such as INSERT, UPDATE, DELETE, and SELECT, as well as CREATE TABLE, ALTER TABLE, and so on. Note that SQL (*http://en.wikipedia.org/wiki/SQL*) is another language altogether, and is not covered in this book. We assume here that you have a very basic knowledge of SQL or can pick it up from the many sources of information that exist. Also note that in sqlite3, you must terminate your SQL statements with a semicolon (;).

- sqlite3 commands. These are commands that are specific to SQLite. They are distinguished by an initial period. You can see the list of all commands by typing .help at the sqlite3> prompt. For now, we'll just use .schema to verify that the schema was created:

```
# sqlite3 timeline.db
SQLite version 3.6.22
Enter ".help" for instructions
Enter SQL statements terminated with a ";"
sqlite> .schema
CREATE TABLE android_metadata (locale TEXT);
CREATE TABLE timeline ( _id integer primary key,created_at integer,
        user text, message text );
```

The last line tells us that our database table timeline indeed was created and looks like we expected, with the columns _id, created_at, message, and user.

New Android developers often execute the `sqlite3 timeline.db` command in a wrong folder, and then wonder why the database table wasn't created. SQLite will not complain if the file you are referring to doesn't exist; it will simply create a brand-new database. So, make sure you are either in the correct folder (*/data/data/com.marakana.android.yamba/ databases/*) when you execute `sqlite3 timeline.db`, or run the command specifying the full path to your file:

```
sqlite3 /data/data/com.marakana.android.yamba/databases/timeline.db
```

Now that we have a way to create and open up our database, we are ready to update the service that will insert the data into the database.

At this point we should be getting the data from the online service as well as inserting that data in the database. We can also verify that the data is indeed in the database by using `sqlite3`. This can be done with the `dump` command at the `sqlite3` prompt:

```
sqlite> .dump
```

Content Providers

Content providers are Android building blocks that can expose data across the boundaries between application sandboxes. As you recall, each application in Android runs in its own process with its own permissions. This means that an application cannot see another app's data. But sometimes you want to share data across applications. This is where content providers become very useful.

Take your contacts, for example. You might have a large database of contacts on your device, which you can view via the Contacts app as well as via the Dialer app. Some devices, such as HTC Android models, might even have multiple versions of the Contacts and Dialer apps. It would not make a lot of sense to have similar data live in multiple databases.

Content providers let you centralize content in one place and have many different applications access it as needed. In the case of the contacts on your phone, there is actually a ContactProvider application that contains a content provider, and other applications access the data via this interface. The interface itself is fairly simple: it has the same `insert()`, `update()`, `delete()`, and `query()` methods we saw in "Databases on Android" on page 175.

Android uses content providers quite a bit internally. In addition to contacts, your settings represent another example, as do all your bookmarks. All the media in the system is also registered with MediaStore, a content provider that dispenses images, music, and videos in your device.

Creating a Content Provider

We're going to create a content provider, `StatusProvider`, that is internal to our app. This is a best practice—your data should be exposed via a provider and nobody but the provider should be concerned with where it comes from, in our case the database. As we said before, the content provider is simply an interface to data, so it is agnostic to the actual data storage.

To create a content provider:

1. Create a new Java class that subclasses the system's `ContentProvider` class.

2. Declare your `CONTENT_URI`.

3. Implement all the unimplemented methods, such as `insert()`, `update()`, `de lete()`, `query()`, `getID()`, and `getType()`.

4. Declare your content provider in the *AndroidManifest.xml* file.

We are going to start by creating a brand-new Java class in the same package as all other classes. Its name will be `StatusProvider`. This class, like any of Android's main building blocks, will subclass an Android framework class, in this case `ContentProvider`.

In Eclipse, select your package, click File → New → Java Class, and enter **StatusProvider**. Then, update the class to subclass `ContentProvider`, and organize the imports (Ctrl-Shift-O) to import the appropriate Java packages. The result should look like this:

```
package com.marakana.android.yamba;

import android.content.ContentProvider;

public class StatusProvider extends ContentProvider {

}
```

Of course, this code is now broken because we need to provide implementations for many of its methods. The easiest way to do that is to click the class name and choose "Add unimplemented methods" from the list of quick fixes. Eclipse will then create stubs, or templates, of the missing methods.

Defining the URI

Objects within a single app share an address space, so they can refer to each other simply by variable names. But objects in different apps don't recognize the different address spaces, so they need some other mechanism to find each other. Android uses a Uniform Resource Identifier (*http://en.wikipedia.org/wiki/Uniform_Resource_Identifier*), a string that identifies a specific resource, to locate a content provider. A URI has three or four parts, shown in Figure 11-1.

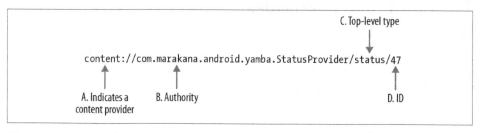

Figure 11-1. Parts of a URI

- Part A, `content://`, is always set to this value. This is written in stone.

- Part B, `com.marakana.android.yamba.StatusProvider`, is the so-called *authority*. It is typically the name of the class, all in lowercase. This authority must match the authority that we specify for this provider when we later declare it in the manifest file.

- Part C, `status`, indicates the type of data that this particular provider provides. It could contain any number of *segments* separated with a slash, including none at all.

- Part D, 47, is an optional ID for the specific item that we are referencing. If not set, the URI will represent the entire set. Number 47 is an arbitrary number picked for this example.

Sometimes you need to refer to the content provider in its entirety, and sometimes to only one of the items of data it contains. If you refer to it in its entirety, you leave off part D; otherwise, you include that part to identify one item within the content provider. Actually, because we have only one table, we do not need part C of the URI.

One way to define the constants for our example is shown in Example 11-4.

Example 11-4. Updated StatusContract.java

```
package com.marakana.android.yamba;

import android.net.Uri;
import android.provider.BaseColumns;

public class StatusContract {

    // DB specific constants
    public static final String DB_NAME = "timeline.db";
    public static final int DB_VERSION = 1;
    public static final String TABLE = "status";

    // Provider specific constants
    // content://com.marakana.android.yamba.StatusProvider/status
    public static final String AUTHORITY = "com.marakana.android.yamba
                              .StatusProvider";
```

```
        public static final Uri CONTENT_URI = Uri.parse("content://" + AUTHORITY
                    + "/" + TABLE);
        public static final int STATUS_ITEM = 1;
        public static final int STATUS_DIR = 2;
        public static final String STATUS_TYPE_ITEM =
    "vnd.android.cursor.item/vnd.com.marakana.android.yamba.provider.status";
        public static final String STATUS_TYPE_DIR =
    "vnd.android.cursor.dir/vnd.com.marakana.android.yamba.provider.status";
        public static final String DEFAULT_SORT = Column.CREATED_AT + " DESC";

        public class Column {
                public static final String ID = BaseColumns._ID;
                public static final String USER = "user";
                public static final String MESSAGE = "message";
                public static final String CREATED_AT = "created_at";
        }
}
```

In "Getting the Data Type" on page 189, we'll explore the reason for two MIME types. We are also going to define the status data object in a class-global variable so that we can refer to it, as Example 11-5 illustrates.

Example 11-5. StatusProvider.java, onCreate()

```
public class StatusProvider extends ContentProvider {
    private static final String TAG = StatusProvider.class.getSimpleName();
        private DbHelper dbHelper;


        @Override
        public boolean onCreate() {
                dbHelper = new DbHelper(getContext());
                Log.d(TAG, "onCreated");
                return true;
        }
    ...
}
```

Getting the Data Type

A content provider must return the MIME type (*http://en.wikipedia.org/wiki/MIME*) of the data it is returning. The MIME type indicates either a single item or all the records for the given URI. Earlier in this chapter we defined the single-record MIME type as vnd.android.cursor.item/vnd.marakana.yamba.status and the directory of all statuses as vnd.android.cursor.dir/vnd.marakana.yamba.status. To let others retrieve the MIME type, we must define the call getType().

The first part of the MIME type is either vnd.android.cursor.item or vnd.an droid.cursor.dir, depending on whether the type represents a specific item or all items for the given URI. The second part, vnd.marakana.yamba.status or vnd.mara kana.yamba.mstatus for our app, is a combination of the constant vnd followed by your company or app name and the actual content type.

As you may recall, the URI can end with a number. If it does, that number is the ID of the specific record. If it doesn't, the URI refers to the entire collection.

Example 11-6 shows the implementation of getType() as well as the getId() helper method that we've already used several times.

Example 11-6. StatusProvider, uri matcher and getType()

```java
public class StatusProvider extends ContentProvider {
    ...
    private static final UriMatcher sURIMatcher = new UriMatcher(
                    UriMatcher.NO_MATCH);
        static {
                sURIMatcher.addURI(StatusContract.AUTHORITY, StatusContract.TABLE,
                        StatusContract.STATUS_DIR);
                sURIMatcher.addURI(StatusContract.AUTHORITY, StatusContract.TABLE
                        + "/#", StatusContract.STATUS_ITEM);
        }

    @Override
        public String getType(Uri uri) {
                switch (sURIMatcher.match(uri)) { ❶
                case StatusContract.STATUS_DIR:
                        Log.d(TAG, "gotType: " + StatusContract.STATUS_TYPE_DIR);
                        return StatusContract.STATUS_TYPE_DIR;
                case StatusContract.STATUS_ITEM:
                        Log.d(TAG, "gotType: " + StatusContract.STATUS_TYPE_ITEM);
                        return StatusContract.STATUS_TYPE_ITEM;
                default:
                        throw new IllegalArgumentException("Illegal uri: " + uri);
                }
        }
    ...
}
```

❶ getType() uses sURIMatcher, an instance of the UriMatcher API class, to determine whether the URI has an ID part. Based on the type of URI we have, getType() returns the appropriate MIME type that we've defined previously in StatusContract.

Although our data is very simple, and we hardly need getType(), this is the best practice for implementing it. You could have a much more complex dataset. Take Android's MediaStore, for example. This is a content provider that contains most of your images, movies, and music—very distinct files.

Inserting Data

To insert a record into a database via the content provider interface, we need to override the `insert()` method. The caller provides the URI of this content provider (without an ID) and the values to be inserted. A successful call to insert the new record returns the ID for that record. We end by returning a new URI concatenating the provider's URI with the ID we just got back, as Example 11-7 illustrates.

Example 11-7. StatusProvider.java, insert()

```java
public class StatusProvider extends ContentProvider {
    ...
    @Override
        public Uri insert(Uri uri, ContentValues values) {
                Uri ret = null;

                // Assert correct uri // ❶
                if (sURIMatcher.match(uri) != StatusContract.STATUS_DIR) {
                        throw new IllegalArgumentException("Illegal uri: " + uri);
                }

                SQLiteDatabase db = dbHelper.getWritableDatabase(); // ❷
                long rowId = db.insertWithOnConflict(StatusContract.TABLE, null,
                                values, SQLiteDatabase.CONFLICT_IGNORE); // ❸

                // Was insert successful?
                if (rowId != -1) { // ❹
                        long id = values.getAsLong(StatusContract.Column.ID);
                        ret = ContentUris.withAppendedId(uri, id); // ❺
                        Log.d(TAG, "inserted uri: " + ret);

                        // Notify that data for this uri has changed
                        getContext().getContentResolver()
                .notifyChange(uri, null); // ❻
                }

                return ret;
        }
    ...
}
```

❶ First, we check whether the URI is valid. The URI that specifies a specific item —for example, `content://com.marakana.android.yamba.StatusProvider/status/47`—is not valid for insert because the ID is not known after the insert happens.

❷ Open the database for writing.

❸ We attempt to insert the values into the database and, upon a successful insert, receive the ID of the new record from the database.

❹   If anything fails during the insert, the database will return −1. We can then throw a runtime exception because this is an error that should never have happened.

❺   If the insert was successful, we use the `ContentUris.withAppendedId()` helper method to craft a new URI containing the ID of the new record appended to the standard provider's URI.

❻   We notify the observers of this content provider that this particular data has changed. This is going to be more obvious in Chapter 12.

Updating Data

To update the data via the Content Provider API, we need:

The URI of the provider
> This may or may not contain an ID. If it does, the ID indicates the specific record that needs to be updated, and we can ignore the selection. If the ID is not specified, it means that we are updating many records and need the selection to indicate which are to be changed.

The values to be updated
> The format of this parameter is a set of name-value pairs that represents column names and new values.

Any selection and arguments that go with it
> These together make up a `WHERE` clause in SQL, selecting the records that will change. The selection and its arguments are omitted when there is an ID, because the ID is enough to select the record that is being updated.

The code that handles both types of update—by ID and by selection—can be seen in Example 11-8.

Example 11-8. StatusProvider.java, update()

```
public class StatusProvider extends ContentProvider {
    ...

    @Override
    public int update(Uri uri, ContentValues values, String selection,
                    String[] selectionArgs) {
        String where;

        switch (sURIMatcher.match(uri)) { // ❶
        case StatusContract.STATUS_DIR:
            // so we count updated rows
            where = selection;  // ❷
            break;
        case StatusContract.STATUS_ITEM:
            long id = ContentUris.parseId(uri);
            where = StatusContract.Column.ID
```

```
                                + "="
                                + id
                                + (TextUtils.isEmpty(selection) ? "" : " and ( "
                                        + selection + " )"); // ❸
                        break;
                default:
                        throw new IllegalArgumentException("Illegal uri: " + uri);
        // ❹
        }

        SQLiteDatabase db = dbHelper.getWritableDatabase();
        int ret = db.update(StatusContract.TABLE, values,
          where, selectionArgs); // ❺

        if(ret>0) { // ❻
                // Notify that data for this URI has changed
                getContext().getContentResolver().notifyChange(uri, null);
        }
        Log.d(TAG, "updated records: " + ret);
        return ret;
    }
    ...
}
```

❶ First, we check the type of URI that was passed in.

❷ If the URI doesn't contain the ID, we don't have much else to worry about.

❸ However, if the URI does have an ID as part of the path, we need to extract it and update our WHERE statement to account for that ID.

❹ We shouldn't be seeing any other type of URI.

❺ Open the database for writing the updates and call update(), passing in this data.

❻ If the update was successful (i.e., the number of affected rows is more than zero), we notify any interested parties that the data has changed.

Deleting Data

Deleting data is similar to updating data. The URI may or may not contain the ID of the particular record to delete, as Example 11-9 illustrates.

Example 11-9. StatusProvider.java, delete()

```
public class StatusProvider extends ContentProvider {
    ...
    @Override
        public int delete(Uri uri, String selection, String[] selectionArgs) {
                String where;

                switch (sURIMatcher.match(uri)) {
```

```
            case StatusContract.STATUS_DIR:
                    // so we count deleted rows
                    where = (selection == null) ? "1" : selection;
                    break;
            case StatusContract.STATUS_ITEM:
                    long id = ContentUris.parseId(uri);
                    where = StatusContract.Column.ID
                            + "="
                            + id
                            + (TextUtils.isEmpty(selection) ? "" : " and ( "
                                    + selection + " )");
                    break;
            default:
                    throw new IllegalArgumentException("Illegal uri: " + uri);
            }

            SQLiteDatabase db = dbHelper.getWritableDatabase();
            int ret = db.delete(StatusContract.TABLE, where, selectionArgs);

            if(ret>0) {
                    // Notify that data for this uri has changed
                    getContext().getContentResolver().notifyChange(uri, null);
            }
            Log.d(TAG, "deleted records: " + ret);
            return ret;
    }
    ...
}
```

Querying Data

Unlike insert(), update(), and delete(), query() returns the actual data and it doesn't modify the underlying dataset. It is analogous to SQL's SELECT statement.

To query the data via a content provider, we override the query() method. This method has a long list of parameters, but usually we just forward most of them to the database call with the same name, as Example 11-10 illustrates.

Example 11-10. StatusProvider.java, query()

```
public class StatusProvider extends ContentProvider {
    ...
    @Override
        public Cursor query(Uri uri, String[] projection, String selection,
                        String[] selectionArgs, String sortOrder) {

            SQLiteQueryBuilder qb = new SQLiteQueryBuilder();    // ❶
            qb.setTables( StatusContract.TABLE );   // ❷

            switch (sURIMatcher.match(uri)) { // ❸
            case StatusContract.STATUS_DIR:
                    break;
```

```
            case StatusContract.STATUS_ITEM:
                    qb.appendWhere(StatusContract.Column.ID + "="
                                    + uri.getLastPathSegment()); // ❹
                    break;
            default:
                    throw new IllegalArgumentException("Illegal uri: " + uri);
            }

            String orderBy = (TextUtils.isEmpty(sortOrder))
                        ? StatusContract.DEFAULT_SORT
                            : sortOrder; // ❺

            SQLiteDatabase db = dbHelper.getReadableDatabase(); // ❻
            Cursor cursor = qb.query(db, projection, selection, selectionArgs,
                    null, null, orderBy); // ❼

            // register for uri changes
            cursor.setNotificationUri(getContext().getContentResolver(), uri);
      // ❽

            Log.d(TAG, "queried records: "+cursor.getCount());
            return cursor; // ❾
    }

    ...
}
```

❶ Here we use `SQLiteQueryBuilder` to make it easier to put together a potentially complex query statement.

❷ Don't forget to specify what table you are working on.

❸ Again, we use the matcher to see what type of the URI we got.

❹ If the URI contains the ID of the record to query, we need to extract that ID and include it in the query. This is where `SQLiteQueryBuilder` makes it easier than building a long string.

❺ Specify the sort order for the returned data, using `default` if sort order hasn't been provided.

❻ We need to open the database, in this case just for reading.

❼ Note that the database call has two additional parameters that correspond to the `GROUPING` and `HAVING` components of a `SELECT` statement in SQL. Because content providers do not support this feature, we simply pass in `null`.

❽ Tell this cursor that it depends on the data as specified by this URI. In other words, when the `insert()`, `update()`, or `delete()` notify the app that the data has changed, this cursor will know that it may want to refresh its data.

❾ Return the actual data in the form of a cursor.

At this point, we have the entire content provider implemented. Example 11-11 shows what the complete code looks like.

Example 11-11. StatusProvider, final

```
package com.marakana.android.yamba;

import android.content.ContentProvider;
import android.content.ContentUris;
import android.content.ContentValues;
import android.content.UriMatcher;
import android.database.Cursor;
import android.database.sqlite.SQLiteDatabase;
import android.database.sqlite.SQLiteQueryBuilder;
import android.net.Uri;
import android.text.TextUtils;
import android.util.Log;

public class StatusProvider extends ContentProvider {
    private static final String TAG = StatusProvider.class.getSimpleName();
        private DbHelper dbHelper;

        private static final UriMatcher sURIMatcher = new UriMatcher(
                    UriMatcher.NO_MATCH);
        static {
                sURIMatcher.addURI(StatusContract.AUTHORITY, StatusContract.TABLE,
                            StatusContract.STATUS_DIR);
                sURIMatcher.addURI(StatusContract.AUTHORITY, StatusContract.TABLE
                            + "/#", StatusContract.STATUS_ITEM);
        }

        @Override
        public boolean onCreate() {
                dbHelper = new DbHelper(getContext());
                Log.d(TAG, "onCreated");
                return true;
        }

        @Override
        public String getType(Uri uri) {
                switch (sURIMatcher.match(uri)) {
                case StatusContract.STATUS_DIR:
                        Log.d(TAG, "gotType: " + StatusContract.STATUS_TYPE_DIR);
                        return StatusContract.STATUS_TYPE_DIR;
                case StatusContract.STATUS_ITEM:
                        Log.d(TAG, "gotType: " + StatusContract.STATUS_TYPE_ITEM);
                        return StatusContract.STATUS_TYPE_ITEM;
                default:
                        throw new IllegalArgumentException("Illegal URI: " + uri);
                }
        }
```

```
@Override
public Uri insert(Uri uri, ContentValues values) {
        Uri ret = null;

        // Assert correct uri
        if (sURIMatcher.match(uri) != StatusContract.STATUS_DIR) {
                throw new IllegalArgumentException("Illegal uri: " + uri);
        }

        SQLiteDatabase db = dbHelper.getWritableDatabase();
        long rowId = db.insertWithOnConflict(StatusContract.TABLE, null,
                        values, SQLiteDatabase.CONFLICT_IGNORE);

        // Was insert successful?
        if (rowId != -1) {
                long id = values.getAsLong(StatusContract.Column.ID);
                ret = ContentUris.withAppendedId(uri, id);
                Log.d(TAG, "inserted uri: " + ret);

                // Notify that data for this uri has changed
                getContext().getContentResolver().notifyChange(uri, null);
        }

        return ret;
}

@Override
public int update(Uri uri, ContentValues values, String selection,
                String[] selectionArgs) {
        String where;

        switch (sURIMatcher.match(uri)) {
        case StatusContract.STATUS_DIR:
                // so we count updated rows
                where = selection;
                break;
        case StatusContract.STATUS_ITEM:
                long id = ContentUris.parseId(uri);
                where = StatusContract.Column.ID
                        + "="
                        + id
                        + (TextUtils.isEmpty(selection) ? "" : " and ( "
                                + selection + " )");
                break;
        default:
                throw new IllegalArgumentException("Illegal uri: " + uri);
        }

        SQLiteDatabase db = dbHelper.getWritableDatabase();
        int ret = db.update(StatusContract.TABLE, values, where,
          selectionArgs);
```

```java
            if(ret>0) {
                    // Notify that data for this uri has changed
                    getContext().getContentResolver().notifyChange(uri, null);
            }
            Log.d(TAG, "updated records: " + ret);
            return ret;
    }


    @Override
    public int delete(Uri uri, String selection, String[] selectionArgs) {
            String where;

            switch (sURIMatcher.match(uri)) {
            case StatusContract.STATUS_DIR:
                    // so we count deleted rows
                    where = (selection == null) ? "1" : selection;
                    break;
            case StatusContract.STATUS_ITEM:
                    long id = ContentUris.parseId(uri);
                    where = StatusContract.Column.ID
                            + "="
                            + id
                            + (TextUtils.isEmpty(selection) ? "" : " and ( "
                                    + selection + " )");
                    break;
            default:
                    throw new IllegalArgumentException("Illegal uri: " + uri);
            }

            SQLiteDatabase db = dbHelper.getWritableDatabase();
            int ret = db.delete(StatusContract.TABLE, where, selectionArgs);

            if(ret>0) {
                    // Notify that data for this uri has changed
                    getContext().getContentResolver().notifyChange(uri, null);
            }
            Log.d(TAG, "deleted records: " + ret);
            return ret;
    }

    // SELECT username, message, created_at FROM status WHERE user='bob' ORDER
    // BY created_at DESC;
    @Override
    public Cursor query(Uri uri, String[] projection, String selection,
                    String[] selectionArgs, String sortOrder) {

            SQLiteQueryBuilder qb = new SQLiteQueryBuilder();
            qb.setTables( StatusContract.TABLE );

            switch (sURIMatcher.match(uri)) {
            case StatusContract.STATUS_DIR:
```

```
                    break;
            case StatusContract.STATUS_ITEM:
                    qb.appendWhere(StatusContract.Column.ID + "="
                                    + uri.getLastPathSegment());
                    break;
            default:
                    throw new IllegalArgumentException("Illegal uri: " + uri);
            }

            String orderBy = (TextUtils.isEmpty(sortOrder))
                        ? StatusContract.DEFAULT_SORT
                             : sortOrder;

            SQLiteDatabase db = dbHelper.getReadableDatabase();
            Cursor cursor = qb.query(db, projection, selection, selectionArgs,
                    null, null, orderBy);

            // register for uri changes
            cursor.setNotificationUri(getContext().getContentResolver(), uri);

            Log.d(TAG, "queried records: "+cursor.getCount());
            return cursor;
    }

}
```

Updating the Android Manifest File

As with any major building block, we want to define our content provider in the *AndroidManifest.xml* file. Notice that in this case the android:authorities property specifies the URI authority permitted to access this content provider. Typically, this authority would be your content provider class—which we use here—or your package:

```
<application>
    ...
        <provider
            android:name="com.marakana.android.yamba.StatusProvider"
            android:authorities="com.marakana.android.yamba.StatusProvider"
            android:exported="false" />

    ...
</application>
```

Notice that we also specify that this provider is not exported at this time. This is for security reasons: we don't want anyone who happens to know the provider's authority to be able to access its data. At this point, we are the only ones using this provider, so android:exported="false" makes sense.

Updating RefreshService

Currently, our content provider is not used by anyone. At the same time, the Refresh-Service goes directly to the database, making it tightly coupled with the storage of the data. What we want to do is refactor this so that the service only talks to the provider, and not to the database, as Example 11-12 illustrates.

Example 11-12. RefreshService, refactored to use StatusProvider instead of the database directly

```java
public class RefreshService extends IntentService {
    ...
    @Override
    protected void onHandleIntent(Intent intent) {
        SharedPreferences prefs = PreferenceManager
                .getDefaultSharedPreferences(this);
        final String username = prefs.getString("username", "");
        final String password = prefs.getString("password", "");

        // Check that username and password are not empty
        if (TextUtils.isEmpty(username) || TextUtils.isEmpty(password)) {
            Toast.makeText(this,
                "Please update your username and password",
                        Toast.LENGTH_LONG).show();
            return;
        }
        Log.d(TAG, "onStarted");

        ContentValues values = new ContentValues();

        YambaClient cloud = new YambaClient(username, password);
        try {
            int count = 0;
            List<Status> timeline = cloud.getTimeline(20);
            for (Status status : timeline) {
                values.clear();
                values.put(StatusContract.Column.ID,
            status.getId());
                values.put(StatusContract.Column.USER,
            status.getUser());
                values.put(StatusContract.Column.MESSAGE,
            status.getMessage());
                values.put(StatusContract.Column.CREATED_AT,
            status.getCreatedAt().getTime());

                    Uri uri = getContentResolver().insert(
                        StatusContract.CONTENT_URI, values); // ❶
                    if (uri != null) {
                        count++;      // ❷
                        Log.d(TAG,
                            String.format("%s: %s", status.getUser(),
                                    status.getMessage()));
```

```
                }
            }

        } catch (YambaClientException e) {
                Log.e(TAG, "Failed to fetch the timeline", e);
                e.printStackTrace();
        }

        return;
    }

    ...
}
```

❶   The only difference is that we now use getContentResolver() from the current context to get the access to content provider's insert(). The actual provider to use is resolved via the URI that we pass: StatusContract.CONTENT_URI, which is registered with the system via the application *AndroidManifest.xml* file. This is how the content resolver knows that it's StatusProvider on the receiving end of this insert() call.

❷   This is a minor addition to start counting how many successful inserts we actually had. For now, we just print out this number.

Summary

At this point, Yamba can pull the statuses of our friends from the cloud and post them into the local database via a content provider. We still don't have a way to view this data, but we can verify that the data is there in the database.

Figure 11-2 illustrates what we have done so far as part of the design outlined earlier in Figure 6-4.

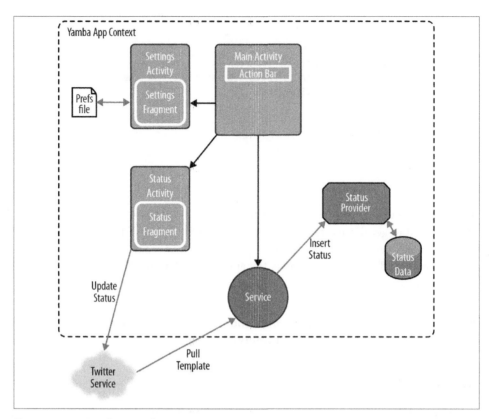

Figure 11-2. Yamba completion

Lists and Adapters

In this chapter, you will learn how to create selection widgets, such as a `ListView`. But this isn't just a chapter about user interface elements. We are deepening our understanding of data from the previous chapter by learning how to read and use data from the status database. At first we'll simply output it to the screen as scrollable text. You will then learn about adapters in order to connect your database directly with the list and create a custom adapter to implement some additional functionality. You will link this new activity with your main activity so that the user can both post and read tweets.

By the end of this chapter, your app will be able to post new tweets, as well as pull them from Twitter, store them in the local database, and let the user read the statuses in a nice and efficient UI. At that point, your app will have three activities and a service.

MainActivity

We're going to create a new `MainActivity`. This activity will become the entry point into the application and for the most part will contain the `TimelineFragment` that will pull the data from the content provider and show it to the user.

Basic MainActivity

As in earlier chapters, the `MainActivity` just loads and inflates the layout, plus handles the Action Bar events, as Example 12-1 illustrates.

Example 12-1. MainActivity

```
package com.marakana.android.yamba;

import android.app.Activity;
import android.content.Intent;
import android.os.Bundle;
import android.view.Menu;
```

```java
import android.view.MenuItem;
import android.widget.Toast;

public class MainActivity extends Activity {

    @Override
    protected void onCreate(Bundle savedInstanceState) {
            super.onCreate(savedInstanceState);
            setContentView(R.layout.activity_main); // ❶
    }

    // Called to lazily initialize the action bar
    @Override
    public boolean onCreateOptionsMenu(Menu menu) { // ❷
            // Inflate the menu items to the action bar.
            getMenuInflater().inflate(R.menu.main, menu);
            return true;
    }

    // Called every time user clicks on an action
    @Override
    public boolean onOptionsItemSelected(MenuItem item) { //❸
            switch (item.getItemId()) {
            case R.id.action_settings:
                    startActivity(new Intent(this, SettingsActivity.class));
                    return true;
            case R.id.action_tweet:
                    startActivity(new Intent(
        "com.marakana.android.yamba.action.tweet"));
                    return true;
            case R.id.action_refresh:
                    startService(new Intent(this, RefreshService.class));
                    return true;
            case R.id.action_purge:
                    int rows = getContentResolver().delete(
        StatusContract.CONTENT_URI, null, null);
                    Toast.makeText(this, "Deleted "+rows+" rows",
        Toast.LENGTH_LONG).show();
                    return true;
            default:
                    return false;
            }
    }
}
```

❶   Inflate the new layout of the main activity.

❷   This is where we load up the action bar menu items.

❸   Finally, we process the menu bar clicks.

Not much of this is new.

However, this activity will load the XML layout resource that will include a Timeline Fragment that will actually process the timeline and display it to the user, as Example 12-2 illustrates.

Example 12-2. The res/layout/activity_main.xml file

```xml
<RelativeLayout xmlns:android="http://schemas.android.com/apk/res/android"
    xmlns:tools="http://schemas.android.com/tools"
    android:layout_width="match_parent"
    android:layout_height="match_parent"
    android:paddingBottom="@dimen/activity_vertical_margin"
    android:paddingLeft="@dimen/activity_horizontal_margin"
    android:paddingRight="@dimen/activity_horizontal_margin"
    android:paddingTop="@dimen/activity_vertical_margin"
    tools:context=".MainActivity" >

    <!-- Timeline Fragment ❶ -->
    <fragment
        android:id="@+id/fragment_timeline"
        android:name="com.marakana.android.yamba.TimelineFragment"
        android:layout_width="match_parent"
        android:layout_height="match_parent"
        android:layout_centerHorizontal="true" />

</RelativeLayout>
```

❶   This is where we reference the TimelineFragment class.

Timeline Fragment

The timeline fragment will eventually display the data, but for now, we'll just set up the new fragment. We'll subclass ListFragment as a special kind of fragment that already contains a list view, as Example 12-3 illustrates.

Example 12-3. Barebones TimelineFragment, no data

```java
...
public class TimelineFragment extends ListFragment {
    private static final String TAG = TimelineFragment.class.getSimpleName(); // ❶
        private static final String[] FROM = { StatusContract.Column.USER,
                StatusContract.Column.MESSAGE, StatusContract.Column.CREATED_AT,
                StatusContract.Column.CREATED_AT }; // ❷
        private static final int[] TO = { R.id.list_item_text_user,
                R.id.list_item_text_message, R.id.list_item_text_created_at,
                R.id.list_item_freshness }; // ❸
        private SimpleCursorAdapter mAdapter; // ❹

    @Override
        public void onActivityCreated(Bundle savedInstanceState) { // ❺
                super.onActivityCreated(savedInstanceState);
```

```
mAdapter = new SimpleCursorAdapter(getActivity(), R.layout.list_item,
    null, FROM, TO, 0); // ❻

setListAdapter(mAdapter); // ❼
    }

}
```

❶   The usual TAG, which we'll use for debugging purposes.

❷   This is the list of column names that map to the database tables, which provide our data.

❸   These are the view IDs to which we'll bind the data. The IDs are from a custom view, R.layout.list_item, which we'll cover next.

❹   Our adapter, to which we'll connect both the data and the view.

❺   onActivityCreated() is called when the activity hosting this fragment has been created.

❻   Here, we create the adapter that glues together the data (right now null) to the custom view R.layout.list_item. It does that by binding the database columns defined by the FROM array to view IDs identified by the TO array.

❼   We finally attach this adapter to the ListView that is already embedded in the ListFragment, of which TimelineFragment is a subclass.

Creating a List Item Layout

Next, let's see what this custom view list_item is. We simply need to define how a single unit of data will be displayed. We'll have a layout with three simple text views displaying who said what and when, as Example 12-4 illustrates.

Example 12-4. R.layout.list_item

```
<?xml version="1.0" encoding="utf-8"?>
<RelativeLayout xmlns:android="http://schemas.android.com/apk/res/android"
    android:id="@+id/list_item_content"
    android:layout_width="match_parent"
    android:layout_height="match_parent"
    android:descendantFocusability="blocksDescendants"
    android:paddingBottom="@dimen/activity_vertical_margin"
    android:paddingLeft="@dimen/activity_horizontal_margin"
    android:paddingRight="@dimen/activity_horizontal_margin"
    android:paddingTop="@dimen/activity_vertical_margin" >

    <!-- ❶ -->
    <TextView
        android:id="@+id/list_item_text_user"
        android:layout_width="wrap_content"
```

```
    android:layout_height="wrap_content"
    android:layout_alignParentLeft="true"
    android:layout_alignParentTop="true"
    android:text="Slashdot"
    android:textAppearance="?android:attr/textAppearanceMedium" />

<!-- ❷ -->
<TextView
    android:id="@+id/list_item_text_created_at"
    android:layout_width="wrap_content"
    android:layout_height="wrap_content"
    android:layout_alignBaseline="@+id/list_item_text_user"
    android:layout_alignBottom="@+id/list_item_text_user"
    android:layout_alignParentRight="true"
    android:text="10 minutes ago"
    android:textAppearance="?android:attr/textAppearanceSmall"
    android:textColor="@android:color/secondary_text_light" />

<!-- ❸ -->
<TextView
    android:id="@+id/list_item_text_message"
    android:layout_width="match_parent"
    android:layout_height="wrap_content"
    android:layout_alignParentLeft="true"
    android:layout_below="@+id/list_item_text_created_at"
    android:autoLink="web"
    android:focusable="false"
    android:linksClickable="true"
    android:text="Android just became the #1 OS on the planet.
                  Take that, Microsoft! http://t.co/123"
    android:textAppearance="?android:attr/textAppearanceSmall" />
```

```
</RelativeLayout>
```

❶ Displays the user who posted this tweet.

❷ This is the data at which the tweet has been posted, displaying in the top-right corner of the tweet.

❸ Finally, the message of the tweet appears below.

About Adapters

A ScrollView will work for a few dozen records. But what if your status database has hundreds or even thousands of records? Waiting to get and print them all would be highly inefficient. The user probably doesn't even care about all of the data anyhow.

To address this issue, Android provides *adapters*. These are a smart way to connect a View with some kind of data source (see Figure 12-1). Typically, your view would be a ListView and the data would come in the form of a Cursor or Array. So adapters come as subclasses of `CursorAdapter` or `ArrayAdapter`. In our case, we have the data in the form of a cursor, so we'll use `CursorAdapter`.

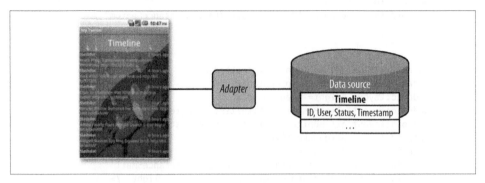

Figure 12-1. Adapter

Loading the Data

Next, we need the data. Our data is already available via the `StatusProvider` we wrote earlier. Now, loading the data from the database could possibly take a long time—we may have a lot of tweets in our timeline. To do that properly, we need to load the data on a separate thread. Once again, we have the problem of not wanting to block the main UI thread.

Android SDK provides a construct called `CursorLoader` designed exactly for this purpose. It consists of an interface and couple of callbacks that are called by the system when the data is ready for us, thus allowing for asynchronous loading, as Example 12-5 illustrates.

Example 12-5. TimelineFragment, with CursorLoader

```
...
public class TimelineFragment extends ListFragment implements
        LoaderCallbacks<Cursor> {    // ❶
        private static final String TAG = TimelineFragment.class.getSimpleName();
        private static final String[] FROM = { StatusContract.Column.USER,
                StatusContract.Column.MESSAGE, StatusContract.Column.CREATED_AT,
                StatusContract.Column.CREATED_AT };
        private static final int[] TO = { R.id.list_item_text_user,
                R.id.list_item_text_message, R.id.list_item_text_created_at,
                R.id.list_item_freshness };
        private static final int LOADER_ID = 42; // ❷
        private SimpleCursorAdapter mAdapter;


        @Override
        public void onActivityCreated(Bundle savedInstanceState) {
```

```
        super.onActivityCreated(savedInstanceState);

        mAdapter = new SimpleCursorAdapter(getActivity(), R.layout.list_item,
                null, FROM, TO, 0);
        mAdapter.setViewBinder(VIEW_BINDER);

        setListAdapter(mAdapter);

        getLoaderManager().initLoader(LOADER_ID, null, this); // ❸
    }

    // --- Loader Callbacks ---

    // Executed on a non-UI thread
    @Override
    public Loader<Cursor> onCreateLoader(int id, Bundle args) { // ❹
        if (id != LOADER_ID)
            return null;
        Log.d(TAG, "onCreateLoader");

        return new CursorLoader(getActivity(), StatusContract.CONTENT_URI,
                null, null, null, StatusContract.DEFAULT_SORT); // ❺
    }

    @Override
    public void onLoadFinished(Loader<Cursor> loader, Cursor cursor) { // ❻
        Log.d(TAG, "onLoadFinished with cursor: " + cursor.getCount());
        mAdapter.swapCursor(cursor); // ❼
    }

    @Override
    public void onLoaderReset(Loader<Cursor> loader) { // ❽
        mAdapter.swapCursor(null);
    }
}
```

❶ We implement LoaderCallbacks<Cursor>, which is the set of callbacks that will be called when the data is available.

❷ This is an arbitrary ID that will help us make sure that the loader calling back is the one we initiated.

❸ When the fragment is created, we initiate the loading of the data. This is now done on a separate thread, not blocking the rest of this method, which as everything else, runs on the UI thread.

❹ onCreateLoader() is where the data is actually loaded. Again, this runs on a worker thread and may take a long time to complete.

❺ A CursorLoader loads the data from the content provider.

❻   Once the data is loaded, the system will call back our code via onLoadFinish ed(), passing in the data.

❼   We update the data that the adapter is using to update the list view. The user finally gets the fresh timeline.

❽   In case the data is stale or unavailable, we remove it from the view.

At this point, MainActivity is complete, but not yet registered with the manifest file. To register it, we'll make MainActivity the entry point into the app, as Example 12-6 illustrates.

Example 12-6. AndroidManifest file

```xml
<?xml version="1.0" encoding="utf-8"?>
<manifest xmlns:android="http://schemas.android.com/apk/res/android"
    package="com.marakana.android.yamba"
    android:versionCode="1"
    android:versionName="1.0" >

    <uses-sdk
        android:minSdkVersion="11"
        android:targetSdkVersion="17" />

    <uses-permission android:name="android.permission.INTERNET" />


    <application
        android:allowBackup="true"
        android:icon="@drawable/ic_launcher"
        android:label="@string/app_name"
        android:theme="@style/AppTheme" >
        <!-- ❶ -->
        <activity
            android:name="com.marakana.android.yamba.StatusActivity"
            android:label="@string/status_update" >
            <intent-filter>
                <action android:name="com.marakana.android.yamba.action.tweet" />

                <category android:name="android.intent.category.DEFAULT" />
            </intent-filter>
        </activity>
        <!-- ❷ -->
        <activity android:name="com.marakana.android.yamba.MainActivity" >
            <intent-filter>
                <action android:name="android.intent.action.MAIN" />

                <category android:name="android.intent.category.LAUNCHER" />
            </intent-filter>
        </activity>
        <activity
            android:name="com.marakana.android.yamba.SettingsActivity"
```

```
            android:label="@string/action_settings" >
        </activity>
        <service android:name="com.marakana.android.yamba.RefreshService" >
        </service>

        <provider
            android:name="com.marakana.android.yamba.StatusProvider"
            android:authorities="com.marakana.android.yamba.StatusProvider"
            android:exported="false" />

    </application>

</manifest>
```

❶ StatusActivity is no longer the main entry point into the app. We moved the main `<intent-filter />` to MainActivity.

❷ `<activity android:name="com.marakana.android.yamba.MainActivity" >` is now the entry point into the app because of the new `<intent-filter />` block.

We can now run the app. However, if we were to run this activity, we'd quickly notice that the timestamp doesn't look quite the way we imagined it.

Remember that we are storing the status creation time in the database as a `long` value representing the number of milliseconds since January 1, 1970. And because that's the value in the database, that's the value we show on the screen as well. This is the standard Unix time (*http://en.wikipedia.org/wiki/Unix_time*), which is very useful for representing actual points in time. But the value is not very meaningful to users. Instead of showing value 1287603266359, it would be much nicer to represent it to the user as "10 minutes ago." This friendly time format is known as *relative time*, and Android provides a method to convert from one format to the other.

The question is where to inject this conversion. As it stands right now, the `SimpleCur sorAdapter` is capable only of mapping straight from a database value to layout view. This doesn't work for our needs, because we need to add some business logic in between the data and the view. To do this, we'll create our own adapter.

Custom Logic via ViewBinder

`ViewBinder` allows us to attach certain business logic to the mapping that the adapter does en route from the cursor to the view.

To attach business logic to an existing `SimpleCursorAdapter`, use its `setViewBind er()` method. We will need to supply the method with an implementation of `ViewBind er`. `ViewBinder` is an interface that specifies `setViewValue()`, where the actual binding of a particular date element to a particular view happens.

Again, we discovered the `setViewBinder()` feature of this `SimpleCursorAdapter` framework class by reading its reference documentation.

 When importing `ViewBinder`, make sure it is `android.widget.Sim pleCursorAdapter.ViewBinder` because there are multiple options.

In our final iteration of `Adapter`, we create a custom `ViewBinder` as a constant and attach it to the stock `SimpleCursorAdapter`, as shown in Example 12-7.

Example 12-7. TimelineFragment with ViewBinder

```
...
public class TimelineFragment extends ListFragment implements
    LoaderCallbacks<Cursor> {
    ...
    @Override
    public void onActivityCreated(Bundle savedInstanceState) {
        super.onActivityCreated(savedInstanceState);
        setEmptyText("Loading data...");

        adapter = new SimpleCursorAdapter(getActivity(), R.layout.row, null,
            FROM, TO, CursorAdapter.FLAG_REGISTER_CONTENT_OBSERVER);
        adapter.setViewBinder(new TimelineViewBinder()); // ❶

        setListAdapter(adapter);

        getLoaderManager().initLoader(0, null, this);
    }

    /** Handles custom binding of data to view. */
    class TimelineViewBinder implements ViewBinder { // ❷

        @Override
        public boolean setViewValue(View view, Cursor cursor,
int columnIndex) { // ❸
            if (view.getId() != R.id.text_created_at) // ❹
                return false;

            // Convert timestamp to relative time
            long timestamp = cursor.getLong(columnIndex); // ❺
            CharSequence relativeTime = DateUtils
                        .getRelativeTimeSpanString(timestamp); // ❻
            ((TextView) view).setText(relativeTime); // ❼

            return true; // ❽
        }
    }
    ...
```

---

❶ We attach a custom `ViewBinder` instance to our stock adapter. `VIEW_BINDER` is defined later in our code.

❷ The actual implementation of a `ViewBinder` instance. Notice that we are implementing it as an inner class. There's no reason for any other class to use it, and thus it shouldn't be exposed to the outside world. Also notice that it is `static final`, meaning that it's a constant.

❸ The only method that we need to provide is `setViewValue()`. This method is called for each data element that needs to be bound to a particular view.

❹ First we check whether this view is the view we care about, i.e., our `TextView` representing when the status was created. If not, we return `false`, which causes the adapter to handle the bind itself in the standard manner. If it is our view, we move on and do the custom bind.

❺ We get the raw timestamp value from the cursor data.

❻ Using the same Android helper method we used in our previous example, `DateUtils.getRelativeTimeSpanString()`, we convert the timestamp to a human-readable format. This is that business logic that we are injecting.

❼ Update the text on the actual view.

❽ Return `true` so that `SimpleCursorAdapter` does not process `bindView()` on this element in its standard way.

Details View

Would it be nice if the user was able to click a specific tweet and get to see its details? We'll do exactly that by creating a new details view. This view will simply display just that one particular selected tweet.

Sometimes this view will need to take an entire screen, such as when the user is on the phone and in portrait mode. Other times, the details view could be right alongside the timeline list.

So this example is going to illustrate the use of fragments again, but more so even the communication between fragments.

Details Fragment

Let's start with the fragment, which represents the reusable piece of UI. Example 12-8 is the code for this fragment. Note that it reuses our R.layout.list_item to render the view (thus not really showing a whole lot more details, but that's not the point).

Example 12-8. DetailsFragment

```
package com.marakana.android.yamba;

import android.app.Fragment;
import android.content.ContentUris;
import android.database.Cursor;
import android.net.Uri;
import android.os.Bundle;
import android.text.format.DateUtils;
import android.view.LayoutInflater;
import android.view.View;
import android.view.ViewGroup;
import android.widget.TextView;

public class DetailsFragment extends Fragment { // ❶
    private TextView textUser, textMessage, textCreatedAt;

    @Override
    public View onCreateView(LayoutInflater inflater, ViewGroup container,
                    Bundle savedInstanceState) {
        View view = inflater.inflate(R.layout.list_item, null, false); // ❷

        textUser = (TextView) view.findViewById(R.id.list_item_text_user);
        textMessage = (TextView) view.findViewById(
                R.id.list_item_text_message);
        textCreatedAt = (TextView) view
                    .findViewById(R.id.list_item_text_created_at);

        return view;
    }

    @Override
    public void onResume() {
        super.onResume();
        long id = getActivity().getIntent().getLongExtra(
                    StatusContract.Column.ID, -1); // ❸

        updateView(id);
    }

    public void updateView(long id) { // ❹
        if (id == -1) {
            textUser.setText("");
            textMessage.setText("");
```

```
                textCreatedAt.setText("");
                return;
        }

        Uri uri = ContentUris.withAppendedId(StatusContract.CONTENT_URI, id);

        Cursor cursor = getActivity().getContentResolver().query(uri, null,
                    null, null, null);
        if (!cursor.moveToFirst())
                return;

        String user = cursor.getString(cursor
                    .getColumnIndex(StatusContract.Column.USER));
        String message = cursor.getString(cursor
                    .getColumnIndex(StatusContract.Column.MESSAGE));
        long createdAt = cursor.getLong(cursor
                    .getColumnIndex(StatusContract.Column.CREATED_AT));

        textUser.setText(user);
        textMessage.setText(message);
        textCreatedAt.setText(DateUtils.getRelativeTimeSpanString(createdAt));
    }
}
```

❶   This is a basic fragment, a reusable piece of UI. We'll attach it to an activity later.

❷   As we said before, we're inflating the R.layout.list_item view that we also use for the list. In another example, you may have a more sophisticated view, but here we want to illustrate something else.

❸   In onResume() we know that this fragment just got redisplayed on the screen. So, we need to update it. To do so, we need to extract the ID for the tweet that we are updating for. We'll assume that whoever requested this fragment to be displayed has passed on that ID to us via the intent that started the activity this fragment is part of. This is similar to a web page for an ecommerce website that you pass in the SKU ID in order to pick the right product to display.

❹   This custom function goes out and pulls the data for the given ID from the content provider, and updates the view of this fragment.

Next, we need to put this fragment somewhere to be visible. There are multiple ways we can do that. One is to make it have the timeline fragment and the details fragment be alongside each other, such as in the case of a larger screen, e.g., tablet or landscape orientation. The other way would be to have each fragment simply be shrink-wrapped in its own minimal activity. Figure 12-2 illustrates both the former and the latter.

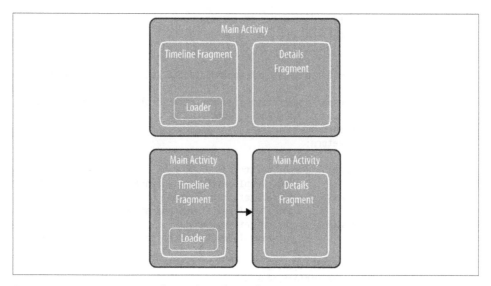

Figure 12-2. Fragments can be packaged together or separate

Details Activity

Let's take the case where we'd just shrink-wrap the details fragment into an activity. Example 12-9 shows what that code could look like.

Example 12-9. DetailsActivity

```
package com.marakana.android.yamba;

import android.os.Bundle;

public class DetailsActivity extends Activity {

    @Override
        protected void onCreate(Bundle savedInstanceState) {
                super.onCreate(savedInstanceState);

                // Check if this activity was created before
                if (savedInstanceState == null) {    // ❶
                        // Create a fragment
                        DetailsFragment fragment = new DetailsFragment(); // ❷
                        getFragmentManager()
                            .beginTransaction()
                            .add(android.R.id.content, fragment,
                                fragment.getClass().getSimpleName()).commit(); // ❸
                }
        }
}
```

❶ We only create the new fragment when onCreate() is called first time around.

❷ Create a new instance of the fragment.

❸ Get the fragment transaction from the manager, and add this fragment to this activity.

Register with the Manifest File

Next, just like with every other activity, we need to register it with the manifest file:

```
<?xml version="1.0" encoding="utf-8"?>
<manifest xmlns:android="http://schemas.android.com/apk/res/android"
    package="com.marakana.android.yamba"
    android:versionCode="1"
    android:versionName="1.0" >

    <application
        android:allowBackup="true"
        android:icon="@drawable/ic_launcher"
        android:label="@string/app_name"
        android:theme="@style/AppTheme" >
        ...

        <!-- ❶ -->
        <activity android:name="com.marakana.android.yamba.DetailsActivity" >
        </activity>

        ...

    </application>
</manifest>
```

❶ We define the DetailsActivity so the system can find it when we try to launch it.

Main Activity, Landscape View

But suppose our activity is viewed on a larger screen? To illustrate that, we'll create a landscape version of the main activity in Example 12-10.

Example 12-10. The res/layout-land/activity_main.xml file

```
<?xml version="1.0" encoding="utf-8"?>
<LinearLayout xmlns:android="http://schemas.android.com/apk/res/android"
    xmlns:tools="http://schemas.android.com/tools"
    android:layout_width="match_parent"
    android:layout_height="match_parent"
    android:orientation="horizontal" >

    <!-- ❶ -->
    <fragment
```

```
    android:id="@+id/fragment_timeline"
    android:name="com.marakana.android.yamba.TimelineFragment"
    android:layout_width="0dp"
    android:layout_height="match_parent"
    android:layout_weight="1"
    tools:layout="@android:layout/list_content" />

<!-- ❷ -->
<fragment
    android:id="@+id/fragment_details"
    android:name="com.marakana.android.yamba.DetailsFragment"
    android:layout_width="0dp"
    android:layout_height="match_parent"
    android:layout_weight="1"
    tools:layout="@layout/list_item" />

</LinearLayout>
```

❶   This is the timeline fragment.

❷   The second fragment is our details fragment. These two fragments will split the horizontal space evenly. This is done via the `android:layout_weight="1"` property, which defines that they both equally yield to one another for the desired width.

Updating TimelineFragment

Finally, we're ready to display `DetailsFragment`. We'll accomplish this by adding an `onListItemClick()` callback to `TimelineFragment`. When the list is clicked, this method will get called and it'll update the details view. Example 12-11 describes the entire `TimelineFragment`.

Example 12-11. TimelineFragment, final version with support for details view

```
package com.marakana.android.yamba;

import android.app.ListFragment;
import android.app.LoaderManager.LoaderCallbacks;
import android.content.CursorLoader;
import android.content.Intent;
import android.content.Loader;
import android.database.Cursor;
import android.os.Bundle;
import android.text.format.DateUtils;
import android.util.Log;
import android.view.View;
import android.widget.ListView;
import android.widget.SimpleCursorAdapter;
import android.widget.SimpleCursorAdapter.ViewBinder;
import android.widget.TextView;
import android.widget.Toast;
```

```java
public class TimelineFragment extends ListFragment implements
        LoaderCallbacks<Cursor> {
    private static final String TAG = TimelineFragment.class.getSimpleName();
    private static final String[] FROM = { StatusContract.Column.USER,
            StatusContract.Column.MESSAGE, StatusContract.Column.CREATED_AT,
            StatusContract.Column.CREATED_AT };
    private static final int[] TO = { R.id.list_item_text_user,
            R.id.list_item_text_message, R.id.list_item_text_created_at,
            R.id.list_item_freshness };
    private static final int LOADER_ID = 42;
    private SimpleCursorAdapter mAdapter;

    private static final ViewBinder VIEW_BINDER = new ViewBinder() {

        @Override
        public boolean setViewValue(View view, Cursor cursor,
    int columnIndex) {
            long timestamp;

            // Custom binding
            switch (view.getId()) {
            case R.id.list_item_text_created_at:
                timestamp = cursor.getLong(columnIndex);
                CharSequence relTime = DateUtils
                        .getRelativeTimeSpanString(timestamp);
                ((TextView) view).setText(relTime);
                return true;
            case R.id.list_item_freshness:
                timestamp = cursor.getLong(columnIndex);
                ((FreshnessView) view).setTimestamp(timestamp);
                return true;
            default:
                return false;
            }
        }
    };

    @Override
    public void onActivityCreated(Bundle savedInstanceState) {
        super.onActivityCreated(savedInstanceState);

        mAdapter = new SimpleCursorAdapter(getActivity(), R.layout.list_item,
                null, FROM, TO, 0);
        mAdapter.setViewBinder(VIEW_BINDER);

        setListAdapter(mAdapter);

        getLoaderManager().initLoader(LOADER_ID, null, this);
    }

    @Override
```

```
public void onListItemClick(ListView l, View v, int position, long id) {
// ❶

        // Get the details fragment
        DetailsFragment fragment = (DetailsFragment) getFragmentManager()
                .findFragmentById(R.id.fragment_details); // ❷

        // Is details fragment visible?
        if (fragment != null && fragment.isVisible()) { // ❸
                fragment.updateView(id); // ❹
        } else {
                startActivity(new Intent(getActivity(), DetailsActivity.class)
                        .putExtra(StatusContract.Column.ID, id)); // ❺
        }
}

// --- Loader Callbacks ---

// Executed on a non-UI thread
@Override
public Loader<Cursor> onCreateLoader(int id, Bundle args) {
        if (id != LOADER_ID)
                return null;
        Log.d(TAG, "onCreateLoader");

        return new CursorLoader(getActivity(), StatusContract.CONTENT_URI,
                null, null, null, StatusContract.DEFAULT_SORT);
}

@Override
public void onLoadFinished(Loader<Cursor> loader, Cursor cursor) {
        // Get the details fragment
        DetailsFragment fragment = (DetailsFragment) getFragmentManager()
                .findFragmentById(R.id.fragment_details);

        // Is details fragment visible?
        if (fragment != null && fragment.isVisible() && cursor.getCount()
== 0) {
                fragment.updateView(-1);
                Toast.makeText(getActivity(), "No data",
    Toast.LENGTH_LONG).show();
        }

        Log.d(TAG, "onLoadFinished with cursor: " + cursor.getCount());
        mAdapter.swapCursor(cursor);
}

@Override
public void onLoaderReset(Loader<Cursor> loader) {
        mAdapter.swapCursor(null);
}
}
```

❶   onListItemClick() is called when an item in the list is clicked on.

❷   We ask the fragment manager for DetailsFragment.

❸   It is quite possible that DetailsFragment is not visible, such as with the portrait orientation of the small phone screen. In that case, DetailsFragment will be null.

❹   If DetailsFragment is not null, it's visible. In that case, we simply call our method updateView() to have the fragment fetch the data from the content provider and update its view.

❺   Otherwise, we launch the details activity, which will do the same once it creates and attaches this fragment to it. Our app now looks like Figure 12-3.

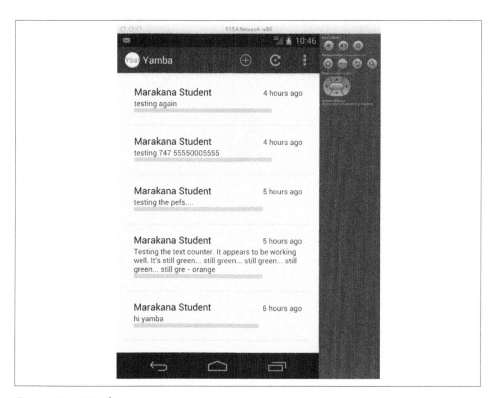

Figure 12-3. Final app

Summary

At this point, Yamba can post a new status as well as list the statuses of our friends. Our application is complete and usable.

Figure 12-4 illustrates what we have done so far as part of the design outlined earlier in Figure 6-4.

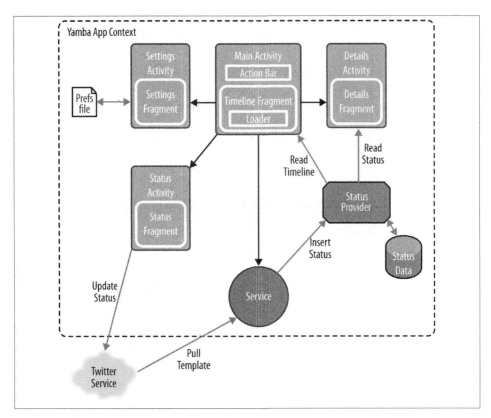

Figure 12-4. Yamba completion

Broadcast Receivers

In this chapter, you will learn about broadcast receivers and when to use them. We'll create a couple of different receivers that illustrate different usage scenarios. First, you'll create a broadcast receiver that will set up the alarms to autostart your refresh timeline service every so often.

Next, you will create a receiver that will be notified when there's a new tweet, and post that notification to the user.

In this chapter, in addition to using broadcast receivers, you will also learn how to take advantage of Android's OS system services.

By the end of this chapter, your app will have most of the functionality that a user would need. The app can send status updates, get friends' timelines, update itself, and start automatically. It works even when the user is not connected to the network (although of course it cannot send or receive new messages).

About Broadcast Receivers

Broadcast receivers are Android's implementation of the publish/subscribe messaging pattern (*http://en.wikipedia.org/wiki/Publish/subscribe*), or more precisely, the observer pattern (*http://en.wikipedia.org/wiki/Observer_pattern*). Applications (known as *publishers*) can generate broadcasts to simply send events without knowing who, if anyone, will get them. Receivers (known as *subscribers*) that want the information subscribe to specific messages via filters. If the message matches a filter, the subscriber is activated (if it's not already running) and notified of the message.

As you may recall from "Broadcast Receivers" on page 72, a `BroadcastReceiver` is a piece of code to which an app subscribes in order to get notified when an action happens. That action is in the form of an *intent broadcast*. When the right intent is fired, the

receiver wakes up and executes. The "wakeup" happens in the form of an onRe
ceive() callback method.

BootReceiver

In our Yamba application, the RefreshService is responsible for periodically updating
the data from the online service. Currently, the user needs to start the service manually,
which she does by starting the application and then clicking the Refresh action bar
button.

It would be much cleaner and simpler if somehow the system automatically started
RefreshService when the device powered up. To do this, we create BootReceiver, a
broadcast receiver that the system will launch when the boot is complete. Example 13-1
sets up our broadcast receiver.

Example 13-1. BootReceiver.java

```
package com.marakana.android.yamba;

import android.content.BroadcastReceiver;
import android.content.Context;
import android.content.Intent;
import android.util.Log;

public class BootReceiver extends BroadcastReceiver { // ❶

  @Override
  public void onReceive(Context context, Intent intent) { // ❷
    context.startService(new Intent(context, RefreshService.class)); // ❸
    Log.d("BootReceiver", "onReceived");
  }

}
```

❶   We create BootReceiver by subclassing BroadcastReceiver, the base class for
     all receivers.

❷   The only method that we need to implement is onReceive(). This method gets
     called when an intent matches this receiver.

❸   We launch an intent to start our Updater service. The system passed us a Con
     text object when it invoked our onReceive() method, and we are expected to
     pass it on to the Updater service. The service doesn't happen to use the Con
     text object for anything, but we'll see an important use for it later.

At this point, we have our BootReceiver. But in order for it to get called—in other
words, in order for the activity to start at boot—we must register it with the system.

Registering the BootReceiver with the Android Manifest File

To register `BootReceiver`, we add it to the manifest file, shown in Example 13-2. We also add an *intent filter* to this file. This intent filter specifies which broadcasts trigger the receiver to become activated.

Example 13-2. AndroidManifest.xml: <application> section

```
...
<receiver android:name=".BootReceiver">
  <intent-filter>
    <action android:name="android.intent.action.BOOT_COMPLETED" />
  </intent-filter>
</receiver>
...
```

In order to get notifications for this particular intent filter, we must also specify that we're using a specific permission it requires, in this case `android.permission.RE CEIVE_BOOT_COMPLETED` (see Example 13-3).

Example 13-3. AndroidManifest.xml: <manifest> section

```
...
<uses-permission android:name="android.permission.RECEIVE_BOOT_COMPLETED" />
...
```

 If we don't specify the permission we require, we simply won't be notified when this event occurs, and we won't have the chance to run our startup code. We won't even know we aren't getting notified, so this is potentially a hard bug to find.

Testing the Boot Receiver

At this point, you can reboot your device. Once it comes back up, your `RefreshSer vice` should be up and running. You can verify this either by looking at the LogCat and verifying that `BootReceiver` did log the "onReceived" message we put in the code.

Alarms and System Services

For now, we got our service started once at the boot time. But what we'd really like to do is have this service be periodically restarted. After all, each trigger of `RefreshSer vice` will pull down the latest timeline data from the cloud. The question is how to create these periodic triggers.

It turns out the Android operating system offers a number of system services that provide useful runtime functionality, and one of them, the Alarm service, has a way to periodically trigger the alarms.

System services, unlike libraries we've dealt with thus far, are always-on-always-running processes. There are around 60+ of these services, such as Alarm, Audio, Camera, Media, Location, Sensors, Telephony, USB, and WiFi, to name a few.

Each service has its own API that is fairly well documented. What is common for all of them is that they are readily available to your app via the context:

```
AlarmManager alarmManager = (AlarmManager) context
                    .getSystemService(Context.ALARM_SERVICE);
```

So, let's see how we can use this particular Alarm service to have our RefreshService started every so often in Example 13-4.

Example 13-4. BootReceiver, final

```
package com.marakana.android.yamba;

import android.app.AlarmManager;
import android.app.PendingIntent;
import android.content.BroadcastReceiver;
import android.content.Context;
import android.content.Intent;
import android.content.SharedPreferences;
import android.preference.PreferenceManager;
import android.util.Log;

public class BootReceiver extends BroadcastReceiver { // ❶
    private static final String TAG = BootReceiver.class.getSimpleName();
        private static final long DEFAULT_INTERVAL =
    AlarmManager.INTERVAL_FIFTEEN_MINUTES; // ❷

        @Override
        public void onReceive(Context context, Intent intent) { // ❸

                SharedPreferences prefs = PreferenceManager
                            .getDefaultSharedPreferences(context);
                long interval = Long.parseLong(prefs.getString("interval",
                            Long.toString(DEFAULT_INTERVAL))); // ❹

                PendingIntent operation = PendingIntent.getService(context, -1,
                            new Intent(context, RefreshService.class),
                            PendingIntent.FLAG_UPDATE_CURRENT); // ❺

                AlarmManager alarmManager = (AlarmManager) context
                            .getSystemService(Context.ALARM_SERVICE); // ❻

                if (interval == 0) { // ❼
                        alarmManager.cancel(operation);
                        Log.d(TAG, "cancelling repeat operation");
                } else {
                        alarmManager.setInexactRepeating(AlarmManager.RTC,
                            System.currentTimeMillis(), interval, operation);
```

```
// ❽
        Log.d(TAG, "setting repeat operation for: " + interval);
    }
    Log.d(TAG, "onReceived");
  }
}
```

❶ The `BootReceiver` is a broadcast receiver, so as with any other building block, we start by subclassing something from the SDK, in this case the `BroadcastRe ceiver` class.

❷ This is going to be our default interval, 15 minutes expressed in milliseconds.

❸ The main callback in broadcast receivers is `onReceive()`, called when the receiver is triggered.

❹ We may have added a property for interval to our settings for the application, alongside the username and password. If not, we'll just use the default value of 15 minutes.

❺ This is where we create our pending intent to be sent by the alarm to trigger the service. Think of the pending intent as an intent plus the action on it, such as start a service.

❻ This is how we get the reference to the system service from the current context.

❼ In case the interval is set to zero, presumably the user doesn't want this service to ever run.

❽ Otherwise, we use the alarm manager's API to repeat this operation every interval, or so.

If you install your app now, and reboot the device, at boot time, the `BootReceiver` will install the alarms that will trigger this sevice to repeat on the given interval.

Broadcasting Intents

In the previous case, the intent that triggered `BootReceiver` was broadcasted by the system. But, you can also broadcast your own intents. Let's say we want to notify the user when there's a new tweet by posting a notification message in the notification bar. To do that, we need to send a broadcast first. We can send that broadcast using the `sendBroadcast()` method in context.

A good place to send the broadcast would be our `RefreshService`—because that's the code that knows there's something new, as Example 13-5 illustrates.

Example 13-5. RefreshService, final

```java
package com.marakana.android.yamba;

import java.util.List;

import android.app.IntentService;
import android.content.ContentValues;
import android.content.Intent;
import android.content.SharedPreferences;
import android.net.Uri;
import android.preference.PreferenceManager;
import android.text.TextUtils;
import android.util.Log;
import android.widget.Toast;

import com.marakana.android.yamba.clientlib.YambaClient;
import com.marakana.android.yamba.clientlib.YambaClient.Status;
import com.marakana.android.yamba.clientlib.YambaClientException;

public class RefreshService extends IntentService {
    private static final String TAG = RefreshService.class.getSimpleName();

    public RefreshService() {
        super(TAG);
    }

    @Override
    public void onCreate() {
        super.onCreate();
        Log.d(TAG, "onCreated");
    }

    // Executes on a worker thread
    @Override
    protected void onHandleIntent(Intent intent) {
        SharedPreferences prefs = PreferenceManager
                .getDefaultSharedPreferences(this);
        final String username = prefs.getString("username", "");
        final String password = prefs.getString("password", "");

        // Check that username and password are not empty
        if (TextUtils.isEmpty(username) || TextUtils.isEmpty(password)) {
            Toast.makeText(this,
        "Please update your username and password",
                    Toast.LENGTH_LONG).show();
            return;
        }
        Log.d(TAG, "onStarted");

        ContentValues values = new ContentValues();

        YambaClient cloud = new YambaClient(username, password);
```

```
            try {
                    int count = 0; // ❶
                    List<Status> timeline = cloud.getTimeline(20);
                    for (Status status : timeline) {
                            values.clear();
                            values.put(StatusContract.Column.ID,
        status.getId());
                            values.put(StatusContract.Column.USER,
        status.getUser());
                            values.put(StatusContract.Column.MESSAGE,
        status.getMessage());
                            values.put(StatusContract.Column.CREATED_AT,
        status.getCreatedAt().getTime());

                            Uri uri = getContentResolver().insert(
                                    StatusContract.CONTENT_URI, values);
                            if (uri != null) {
                                    count++; // ❷
                                    Log.d(TAG,
                                        String.format("%s: %s", status.getUser(),
                                            status.getMessage()));
                            }
                    }

                    if (count > 0) {
                            sendBroadcast(new Intent(
                                "com.marakana.android.yamba.action.NEW_STATUSES")
                    .putExtra("count", count)); // ❸
                    }

            } catch (YambaClientException e) {
                    Log.e(TAG, "Failed to fetch the timeline", e);
                    e.printStackTrace();
            }

            return;
    }

    @Override
    public void onDestroy() {
            super.onDestroy();
            Log.d(TAG, "onDestroyed");
    }
}
```

❶ Initialize the count of new tweets to zero.

❷ If there was a new tweet, increment the counter.

❸ In case we have at least one new tweet, let's use sendBroadcast() to send a broadcast to whoever cares about that.

Notification Receiver

Now we can create a receiver that will receive this broadcast from us, and use another system service to post a notification to the user, as Example 13-6 illustrates.

Example 13-6. NotificationReceiver

```
package com.marakana.android.yamba;

import android.app.Notification;
import android.app.NotificationManager;
import android.app.PendingIntent;
import android.content.BroadcastReceiver;
import android.content.Context;
import android.content.Intent;

public class NotificationReceiver extends BroadcastReceiver { // ❶
    public static final int NOTIFICATION_ID = 42;

        @Override
        public void onReceive(Context context, Intent intent) { // ❷
                NotificationManager notificationManager = (NotificationManager)
        context
                        .getSystemService(Context.NOTIFICATION_SERVICE); // ❸

                int count = intent.getIntExtra("count", 0); // ❹

                PendingIntent operation = PendingIntent.getActivity(context, -1,
                        new Intent(context, MainActivity.class),
                        PendingIntent.FLAG_ONE_SHOT); // ❺

                Notification notification = new Notification.Builder(context)
                        .setContentTitle("New tweets!")
                        .setContentText("You've got " + count + " new tweets")
                        .setSmallIcon(android.R.drawable.sym_action_email)
                        .setContentIntent(operation)
                        .setAutoCancel(true)
                        .getNotification(); // ❻
                notificationManager.notify(NOTIFICATION_ID, notification); // ❼
        }
}
```

❶ As before, each receiver subclasses from BroadcastReceiver.

❷ Again, the magic happens in onReceive().

❸ Similarly to the Alarm service, we get the notification service by calling getSystemService() from the context.

❹ The intent that triggered this receiver was posted in RefreshService, and in there we attached a primitive integer representing the count of new tweets. Here, we extract it from that receiving intent.

❺ We create a pending operation—in other words, what will happen once a user clicks this specific notification. In this case, we launch `MainActivity` so the user can quickly read new tweets.

❻ To post a notification, first we need to build it. This code uses the `Notifica tion.Builder` class to help build a notification with the minimal set of bells and whistles.

❼ Finally, we post this notification to the notification manager.

Summary

Yamba is now complete and ready for prime time. Our application can now send status updates to our online service, get the latest statuses from our friends, start automatically at boot time, and refresh the display when a new status is received.

Figure 13-1 illustrates what we have done so far as part of the design outlined earlier in Figure 6-4.

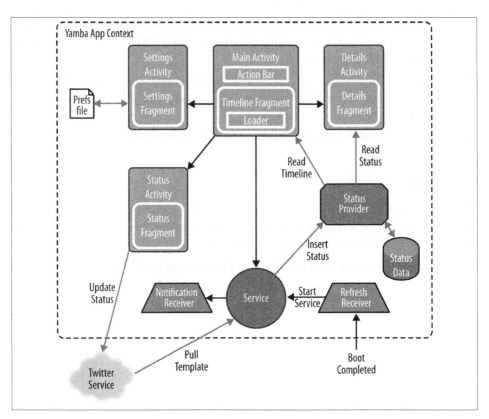

Figure 13-1. Yamba completion

App Widgets

In Android, the idea of showing mini application views embedded in other applications, the most common case being that of the home screen, is a very important and useful feature. These are called app widgets, or widgets for short. These widgets not only provide a small window into an easily accessible view, but also can receive updates and thus provide a more dynamic experience to your application.

Using Content Providers Through Widgets

As mentioned before, content providers make the most sense when you want to expose the data to other applications. It is a good practice to always think of your application as part of a larger Android ecosystem and, as such, a potential provider of useful data to other applications.

To demonstrate how content providers can be useful, we'll create a home screen widget. We're not using the term *widget* here as a synonym for Android's View class, but as a useful embedded service offered by the home screen.

Android typically ships with a few home screen widgets. You can access them by going to your home screen, long-pressing on it to pull up an Add to Home Screen dialog, and choosing Widgets. Widgets that come with Android include Alarm Clock, Picture Frame, Power Controls, Music, and Search. Our goal is to create our own Yamba widget that the user will be able to add to the home screen.

The Yamba widget will be simple, displaying just the latest status update. To create it, we'll make a new YambaWidget class that subclasses AppWidgetProviderInfo. We'll also have to register the widget with the manifest file.

Implementing the YambaWidget Class

YambaWidget is the main class for our widget. It is a subclass of AppWidgetProvider, a special system class that makes widgets. This class itself is a subclass of BroadcastRe ceiver, so our Yamba widget is a broadcast receiver automatically. Basically, whenever our widget is updated, deleted, enabled, or disabled, we'll get a broadcast intent with that information. So this class inherits the onUpdate(), onDeleted(), onEnabled(), onDisabled(), and onReceive() callbacks. We can override any of these, but typically we care mostly about the updates and general broadcasts we receive.

Now that we understand the overall design of the widget framework, Example 14-1 shows how we implement it.

Example 14-1. YambaWidget.java

```
package com.marakana.android.yamba;
package com.marakana.android.yamba;

import android.app.PendingIntent;
import android.appwidget.AppWidgetManager;
import android.appwidget.AppWidgetProvider;
import android.content.ComponentName;
import android.content.Context;
import android.content.Intent;
import android.database.Cursor;
import android.text.format.DateUtils;
import android.util.Log;
import android.widget.RemoteViews;

public class YambaWidget extends AppWidgetProvider { // ❶
    private static final String TAG = YambaWidget.class.getSimpleName();

    @Override
    public void onUpdate(Context context, AppWidgetManager appWidgetManager,
            int[] appWidgetIds) { // ❷
        Log.d(TAG, "onUpdate");

        // Get the latest tweet
        Cursor cursor = context.getContentResolver().query(
                StatusContract.CONTENT_URI, null, null, null,
                StatusContract.DEFAULT_SORT); // ❸

        if (!cursor.moveToFirst()) // ❹
                return;

// ❺
        String user = cursor.getString(cursor
                .getColumnIndex(StatusContract.Column.USER));
        String message = cursor.getString(cursor
                .getColumnIndex(StatusContract.Column.MESSAGE));
        long createdAt = cursor.getLong(cursor
```

```
                .getColumnIndex(StatusContract.Column.CREATED_AT));

        PendingIntent operation = PendingIntent.getActivity(context, -1,
                new Intent(context, MainActivity.class),
                PendingIntent.FLAG_UPDATE_CURRENT);

        // Loop through all the instances of YambaWidget
        for (int appWidgetId : appWidgetIds) { // ❻

                // Update the view
                RemoteViews view = new RemoteViews(context.getPackageName(),
                        R.layout.widget); // ❼

                // Update the remote view ❽
                view.setTextViewText(R.id.list_item_text_user, user);
                view.setTextViewText(R.id.list_item_text_message, message);
                view.setTextViewText(R.id.list_item_text_created_at,
                        DateUtils.getRelativeTimeSpanString(createdAt));
                view.setOnClickPendingIntent(R.id.list_item_text_user,
        operation);
                view.setOnClickPendingIntent(R.id.list_item_text_message,
        operation);

                // Update the widget
                appWidgetManager.updateAppWidget(appWidgetId, view); // ❾
        }

    }


    @Override
    public void onReceive(Context context, Intent intent) { // ❿
            super.onReceive(context, intent);
            AppWidgetManager appWidgetManager = AppWidgetManager
                    .getInstance(context); // ⓫
            this.onUpdate(context, appWidgetManager, appWidgetManager
                    .getAppWidgetIds(new ComponentName(context,
        YambaWidget.class))); // ⓬

    }
}
```

❶   As mentioned before, our widget is a subclass of `AppWidgetProvider`, which
    itself is a `BroadcastReceiver`.

❷   This method is called whenever our widget is to be updated, so it's where we'll
    implement the main functionality of the widget. When we register the widget
    with the system in the manifest file later, we'll specify the update frequency we'd
    like. In our case, this method will be called about every 30 minutes.

❸   We finally get to use our content provider. The whole purpose of the widget in this chapter is to illustrate how to use the `StatusProvider` that we created earlier. As you saw earlier when we implemented the content provider, its API is quite similar to the SQLite database API. The main difference is that instead of passing a table name to a database object, we're passing a content URI to the `ContentResolver`. We still get back the very same `Cursor` object as we did with databases in "Databases on Android" on page 175.

❹   In this particular example, we care only about the very latest status update from the online service. So we position the cursor to the first element. If one exists, it's our latest status update.

❺   In the next few of lines of code, we extract data from the `Cursor` object and store it in local variables.

❻   Because the user could have multiple Yamba widgets installed, we need to loop through them and update them all. We don't particularly care about the specific `appWidgetId` because we're doing identical work to update every instance of the Yamba widget. The `appWidgetId` becomes an opaque handle we use to access each widget in turn.

❼   The actual view representing our widget is in another process. To be precise, our widget is running inside the Home application, which acts as its host and is the process we are updating. Hence the `RemoteViews` constructor. The `RemoteViews` framework is a special shared memory system designed specifically for widgets.

❽   Once we have the reference to our widget views' Java memory space in another process, we can update those views. In this case, we're setting the status data in the row that represents our widget.

❾   Once we update the remote views, the `AppWidgetManager` call to `updateAppWidget()` actually posts a message telling the system to update our widget. This will happen asynchronously, but shortly after `onUpdate()` completes.

❿   The call to `onReceive()` is not necessary in a typical widget. But because a widget is a broadcast receiver, and because our Updater service does send a broadcast when we get a new status update, this method is a good opportunity to invoke `onUpdate()` and get the latest status data updated on the widget.

⓫   If it was, we get the instance of `AppWidgetManager` for this context.

⓬   We then invoke `onUpdate()`.

At this point, we have coded the Yamba widget, and as a receiver, it will be notified periodically or when there are new updates, and it will loop through all instances of this widget on the home screen and update them.

Next, we need to set up the layout for our widget.

Creating the XML Layout

The layout for the widget is fairly straightforward. In Example 14-2, we just include it along with a little title and an icon to make it look good on the home screen.

Example 14-2. The res/layout/widget.xml file

```xml
<?xml version="1.0" encoding="utf-8"?>
<RelativeLayout xmlns:android="http://schemas.android.com/apk/res/android"
    android:id="@+id/list_item_content"
    android:layout_width="match_parent"
    android:layout_height="match_parent"
    android:background="@android:drawable/dialog_holo_dark_frame"
    android:descendantFocusability="blocksDescendants"
    android:paddingBottom="@dimen/activity_vertical_margin"
    android:paddingLeft="@dimen/activity_horizontal_margin"
    android:paddingRight="@dimen/activity_horizontal_margin"
    android:paddingTop="@dimen/activity_vertical_margin" >

    <TextView
        android:id="@+id/list_item_text_user"
        android:layout_width="wrap_content"
        android:layout_height="wrap_content"
        android:layout_alignParentLeft="true"
        android:layout_alignParentTop="true"
        android:text="Slashdot"
        android:textAppearance="?android:attr/textAppearanceMedium" />

    <TextView
        android:id="@+id/list_item_text_created_at"
        android:layout_width="wrap_content"
        android:layout_height="wrap_content"
        android:layout_alignBaseline="@+id/list_item_text_user"
        android:layout_alignBottom="@+id/list_item_text_user"
        android:layout_alignParentRight="true"
        android:text="10 minutes ago"
        android:textAppearance="?android:attr/textAppearanceSmall" />

    <TextView
        android:id="@+id/list_item_text_message"
        android:layout_width="match_parent"
        android:layout_height="wrap_content"
        android:layout_alignParentLeft="true"
        android:layout_below="@+id/list_item_text_created_at"
        android:autoLink="web"
        android:focusable="false"
        android:linksClickable="true"
        android:text="Andriod just became the #1 OS on the planet.
                Take that, Microsoft! http://t.co/123"
        android:textAppearance="?android:attr/textAppearanceSmall" />
```

```
</RelativeLayout>
```

This layout is simple enough, but it does the job for our particular needs. Next, we need to define some basic information about this widget and its behavior.

Creating the AppWidgetProviderInfo File

The XML file shown in Example 14-3 is responsible for describing the widget. It typically specifies which layout this widget uses, how frequently it should be updated by the system, and its size.

Example 14-3. The res/xml/yamba_widget.xml file

```xml
<?xml version="1.0" encoding="utf-8"?>
<appwidget-provider xmlns:android="http://schemas.android.com/apk/res/android"
    android:minHeight="40dp"
    android:minWidth="250dp"
    android:resizeMode="none"
    android:updatePeriodMillis="1800000"
    android:widgetCategory="home_screen|keyguard" >
</appwidget-provider>
```

In this case we specify that we'd like to have our widget updated every 30 minutes or so (1,800,000 milliseconds). Here, we also specify the layout to use, the title of this widget, and its size.

Updating the Manifest File

Finally, we need to update the manifest file and register the widget:

```xml
    ...
    <application .../>
      ...
        <receiver
            android:name="com.marakana.android.yamba.YambaWidget"
            android:exported="false"  >
            <intent-filter>
                <action android:name=
                "com.marakana.android.yamba.action.NEW_STATUSES" />
            </intent-filter>
            <intent-filter>
                <action android:name=
                "android.appwidget.action.APPWIDGET_UPDATE" />
            </intent-filter>

            <meta-data
                android:name="android.appwidget.provider"
                android:resource="@xml/yamba_widget" />
        </receiver>
```

```
...
</application>
...
```

Notice that the widget is a receiver, as we mentioned before. So, just like other broadcast receivers, we declare it within a `<receiver>` tag inside an `<application>` element. It is important to register this receiver to receive `ACTION_APPWIDGET_UPDATE` updates. We do that via the `<intent-filter>`. The `<meta-data>` specifies the meta information for this widget in the *yamba_widget_info* XML file described in the previous section.

That's it. We now have the widget and are ready to test it.

Testing the Widget

To test this widget, install your latest application on the device. Next, go to the home screen, long-press it, and click the Widgets choice. You should be able to navigate to the Yamba widget at this point. After adding it to the home screen, the widget should display the latest status update.

If your Updater service is running, the latest updates should show up on the home screen. This means your widget is running properly.

Summary

At this point, the Yamba app is complete. Congratulations! You are ready to fine-tune it, customize it, and publish it to the market.

Figure 14-1 illustrates what we have done so far as part of the design outlined earlier in Figure 6-4.

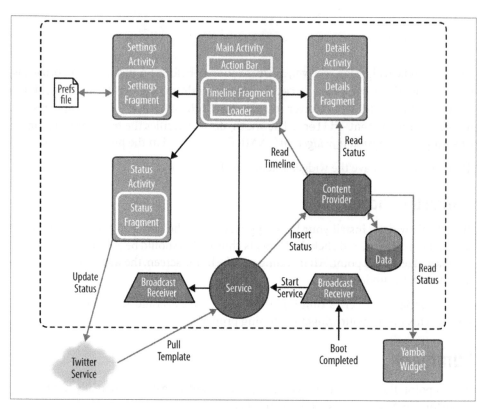

Figure 14-1. Yamba completion

Networking and Web Overview

Networking is one of the fundamental tasks of mobile development. In today's world, the power of the smartphone lies not so much in its computational abilities as in its connections to the greater collection of servers and clients that make up the Internet. Understanding the role of networking within the Android development environment is critical. With this in mind, this chapter will cover a common form of networking: sending web data over HTTP.

Quick Example

Let us do a quick simple example of an HTTP network connection to give an idea of what we are going to cover. First, copy the code in Example 15-1 to a file called *QuickHttpExample.java* and run it (Example 15-2). This will produce the output as shown in Example 15-3 (we truncated the output because it is very long). What this example does is creates an HTTP connection to *http://yamba.marakana.com/api/statuses/public_timeline.json* (if you copy this URL into a browser you will see some very long data that is similar to the long output in the output of the program). It then proceeds to check the response status (`getResponseCode()`), read in the server's output, and write that out to the system console. It terminates the connection after it is done (`disconnect()`).

Example 15-1. Quick example with HttpUrlConnection

```
package com.marakana.examples;

import java.io.BufferedReader;
import java.io.InputStreamReader;
import java.net.URL;
import java.net.HttpURLConnection;

public class QuickHttpExample {
    public static void main(String[] args) {
```

```
HttpURLConnection urlConnection = null;
try {
  URL url =
    new URL(
      "http://yamba.marakana.com/api/statuses/public_timeline.json");

  urlConnection = (HttpURLConnection) url.openConnection();

  int statusCode = urlConnection.getResponseCode();
  System.out.println("Response Code: "+statusCode);

  BufferedReader in = new BufferedReader(
      new InputStreamReader(urlConnection.getInputStream()));

  String textline = null;

  while((textline = in.readLine()) != null) {
      System.out.println(textline);
  }
} catch (Exception e) {
      e.printStackTrace();
} finally {
      if(urlConnection != null) urlConnection.disconnect();
}
  }
}
```

Example 15-2. Compiling and running the quick example

```
javac -d . QuickHttpExample.java

java -cp . com.marakana.examples.QuickHttpExample
```

Example 15-3. Output of quick example (we truncated the text because it is very long)

```
Response Code: 200
[{"text":"as","truncated":false,"created_at".....
```

So now that we have a quick example that runs at the command line, let's take a dive into some of the details around networking and HTTP libraries, and then proceed to tackle some Android-specific methods of network communication.

Networking Basics

Computer networks rely on a set of standards to enable communication between endpoints. Most of this communication using the Internet is based on the Internet Protocol (IP), which describes the addressing and formatting of the data being sent between endpoints. The network socket (in this case an "Internet Socket") is the interface that sits on top of this standard. By using a network socket the developer can ship data based on a standard protocol or communication process. In the case of the World Wide Web

(WWW) we are speaking of HTTP (HyperText Transfer Protocol). HTTP relies on a connection-oriented socket, which is a network socket that uses the Transmission Control Protocol (TCP) to describe how the communication is to occur (initiating a connection, data transmission, signaling transmission and acknowledgment, and terminating a connection).

So what does this all mean? When you browse the Web, you use a web browser with which you type in a URL (uniform resource locator) that describes a destination for your browser to point to. This destination is in the form of *scheme://server/request-uri* (generally the "scheme" portion is "http" or "https"). The browser then executes the communication process, which opens an HTTP socket connection to the server (the *request*) and upon receiving the data from the server (the *response*) proceeds to interpret it and then display the interpretation to you. The connection is generally terminated once the data is processed.

The request sent via the browser uses a small set of methods, depending on the type of request that is being sent. The most common methods are GET and POST. The HTTP/1.0 standard defines the GET, POST, and HEAD methods. HTTP/1.1 adds GET, POST, HEAD, OPTIONS, PUT, DELETE, TRACE, and CONNECT. The methods are used as follows:

GET
> Data retrieval

POST
> Passes data with the attached payload (used often with submission of web forms)

HEAD
> Like the GET method, but requests only metadata, not the data itself

OPTIONS
> Requests the methods supported by the server

PUT
> Puts data at the specified URI

DELETE
> Deletes data dictated by the specified URI

TRACE
> Echoes back the request so the client originator may see what occurs to it

CONNECT
> Converts the requested connection to a transparent TCP/IP tunnel for things such as proxying an encrypted communication

Aside from the request method, the requester generally supplies some metadata about itself, as well as the payload of data if it is needed (such as when using POST and PUT). A typical GET request, for instance, sends over things like User-Agent, which describes what the client is (IE 8 on Windows 7, etc.) and Accept-Encoding, which can help specify compression schemes for the data being transmitted. The request and metadata is sent in simple, plain text lines, such as in Example 15-4.

Example 15-4. Sample GET

```
GET / HTTP/1.0
User-Agent: Mozilla/4.0 (compatible; MSIE 7.0; Windows NT 6.0)
Accept: */*
Host: hostname:port
Accept-Encoding: gzip
```

The response from the server includes header information describing the response, followed by the data (should a data payload be requested and no error occur). The header information includes the content length in bytes, the content type (text, binary media, etc.), and the status code of the response. The response code also follows the HTTP specification with a three-digit number code. Codes of the form 2xx indicate success, 3xx indicate redirection, 4xx indicate some sort of client error (404 being the most famous), and 5xx indicate a server error.

Example 15-5 shows an example of a successful response.

Example 15-5. Sample response

```
HTTP/1.0 200 OK
Server: Apache/2.2.16
Date: Mon, 01 Apr 2013 19:14:30 GMT
Last-modified: Mon, 17 Dec 2012 09:26:13 GMT
Content-length: 6372
Content-type: text/html

<!DOCTYPE HTML PUBLIC "-//W3C//DTD HTML 3.2 Final//EN">
<HTML>
<HEAD>
<TITLE>
HTML DOCUMENT TITLE
</TITLE>
</HEAD>
<BODY>
HTML DOCUMENT DATA
</BODY>
</HTML>
```

HTTP API

So, given all of the communication protocols and processes we've described, thankfully Android has provided within its APIs several classes and methods of generating an HTTP connection and handling the subsequent responses easily. The first of these is the Apache HTTP client classes: `DefaultHttpClient` and `AndroidHttpClient`. These client classes are suitable for web browsers and encompass quite a bit of functionality. Unfortunately, due to the classes' complexity, the Android team at Google is not actively working on them, so you can use them only on the Eclair (2.0) and Froyo (2.1/2.2) versions of Android. For Gingerbread (2.3) or above, another class (`HttpUrlConnection`) is the better choice.

 Whenever you wish to have your application connect out to the Internet, you need to include the `android.permission.INTERNET` permission in the *Manifest.xml* file, as follows:

```
<uses-permission android:name="android.permission.INTERNET">
```

Apache HTTP Client

The Apache HTTP Client (found in the package `org.apache.http.impl.client.DefaultHttpClient`) and the Android HTTP Client (found in `android.net.http.AndroidHttpClient`) are convenience classes to handle connecting and retrieving data via HTTP. Both `DefaultHttpClient` and `AndroidHttpClient` implement the `org.apache.http.client.HttpClient` interface. (Be aware that, even though the class overview for `AndroidHttpClient` states that it is an implementation of `DefaultHttpClient`, that is not the case). There are some differences however. `DefaultHttpClient` has been around since Android 1.0 (see Example 15-6), whereas `AndroidHttpClient` was introduced in Froyo (2.2) (see Example 15-7). Also, `AndroidHttpClient` allows for SSL management, easy methods to specify the `UserAgent` string, and other nice utility methods to set the header information for a request. Instance generation is differs in each class's case: with `DefaultHttpClient` you create an instance via one of its constructors, whereas `AndroidHttpClient` has a factory method called `newInstance()` that takes in `UserAgent` information (which may be null if no `UserAgent` string is needed).

Example 15-6. DefaultHttpClient instantiation

```
DefaultHttpClient client = new DefaultHttpClient();
```

Example 15-7. AndroidHttpClient instantiation

```
AndroidHttpClient client = AndroidHttpClient.newInstance(null);
```

In either class's case you may wish to change some client parameters such as the connection timeout or the socket timeout. A connection timeout is the amount of time the client waits for the server to respond, whereas a socket timeout is the amount of time the client waits when data is coming in and the flow of data is interrupted. `DefaultHttpClient`, by default, does not set the connection timeout or socket timeout (their values are 0), which means the client will not time out at all. `AndroidHttpClient` sets both timeouts to 60 seconds by default. In order to change these parameters, use the `HttpPar ams` and `HttpConnectionParams` classes (see Example 15-8 and Example 15-9).

Example 15-8. DefaultHttpClient instantiation with parameters

```
HttpParams params = new BasicHttpParams();
  // setting connection timeout to 10000 ms (10 seconds)
HttpConnectionParams.setConnectionTimeout(params, 10000);
  // setting socket timeout to 10000 ms (10 seconds)
HttpConnectionParams.setSoTimeout(params, 10000);
DefaultHttpClient client = new DefaultHttpClient(params);
```

Example 15-9. AndroidHttpClient instantiation with parameters

```
AndroidHttpClient client = AndroidHttpClient.newInstance(null);
HttpParams myParams = client.getParams();
  // setting connection timeout to 10000 ms (10 seconds)
HttpConnectionParams.setConnectionTimeout(myParams, 10000);
  // setting socket timeout to 10000 ms (10 seconds)
HttpConnectionParams.setSoTimeout(myParams, 10000);
```

After a client (using either client class) is generated and configured, an HTTP request may be made. To do this, generate a request object (a class implementing `HttpUriRe quest`) and call the client's `execute()` method. Let's look at two examples: a `GET` request and a `POST` request.

In the case of a `GET` request, you need just specify the URI you want to retrieve and execute the `GET` request (see Example 15-10). The URI is provided as an input to the `HttpGet` class constructor. The return value is the `HttpResponse`.

Example 15-10. GET request

```
HttpClient client = new DefaultHttpClient();

String getURL = "http://www.someserver.com/getrequest;
HttpGet get = new HttpGet(getURL);

HttpResponse responseGet = client.execute(get);
```

In the case of a `POST` request, you add your data as part of the request payload. This is done in one of two ways: a key-value paired list or a binary multipart data stream. To pass key-value pairs, create a list of `NameValuePairs` and pass it into the `HttpPost` instance (see Example 15-11). Finish by executing this instance.

Example 15-11. POST request passing a key-value list

```
HttpClient client = new DefaultHttpClient();

String postURL = "http://www.someserver.com";
HttpPost post = new HttpPost(postURL);
```

```
List<NameValuePair> params = new ArrayList<NameValuePair>();
params.add(new BasicNameValuePair("key1", "value1"));
params.add(new BasicNameValuePair("key2", "value2"));
UrlEncodedFormEntity ent = new UrlEncodedFormEntity(params,HTTP.UTF_8);
post.setEntity(ent);

HttpResponse responsePOST = client.execute(post);
```

To pass a multipart file (binary or text), add it as a "part" to the HttpPost instance, as Example 15-12 illustrates.

Example 15-12. POST request passing a multipart file

```
HttpClient client = new DefaultHttpClient();

File file = new File("somefile.txt");

String postURL = "http://www.someserver.com";
HttpPost post = new HttpPost(postURL);
FileBody bin = new FileBody(file);
MultipartEntity reqEntity =
  new MultipartEntity(HttpMultipartMode.BROWSER_COMPATIBLE);
reqEntity.addPart("someFile", bin);
post.setEntity(reqEntity);

HttpResponse response = client.execute(post);
```

After you execute your code, and the server handles it successfully, the library returns an HttpResponse instance. The communication happens synchronously—that is, the *execute* call does not return until the server responds. So at this point in the code, the client thread is now waiting for the response. This would block the device's user from doing anything if you ran the code in the main UI thread (such as running this directly as part of an activity), so using the main thread is not allowed. The issue can be addressed by using AsyncTask or AsyncTaskLoader, which we cover in "Networking in the Background using AsyncTask and AsyncTaskLoader" on page 251. Once the HttpResponse is returned, the status code can be retrieved, as well as the underlying data payload. The payload is returned as an InputStream and thus must be managed like any other I/O stream of data, as Example 15-13 illustrates.

Example 15-13. The full deal using AndroidHttpClient: doing a POST and handling the response

```
AndroidHttpClient client = AndroidHttpClient.newInstance(null);
HttpParams myParams = client.getParams();
HttpConnectionParams.setConnectionTimeout(myParams, 10000);
HttpConnectionParams.setSoTimeout(myParams, 10000);

String postURL = "http://www.someserver.com";
HttpPost post = new HttpPost(postURL);
List<NameValuePair> params = new ArrayList<NameValuePair>();
```

```
params.add(new BasicNameValuePair("key1", "value1"));
params.add(new BasicNameValuePair("key2", "value2"));
UrlEncodedFormEntity ent = new UrlEncodedFormEntity(params,HTTP.UTF_8);
post.setEntity(ent);

HttpResponse responsePOST = client.execute(post);

HttpEntity resEntity = response.getEntity();

if (resEntity != null) {
        System.out.println("Response Code: "+
                        resEntity.getStatusLine().getStatusCode());
        System.out.println("Response Content Length: "+
                        resEntity.getContentLength());
        System.out.println("Response Content Encoding: "+
                        resEntity.getContentEncoding().getValue());
        InputStream is = resEntity.getContent();
        //  then read in the InputStream and Do Something
        is.close();
}
```

HttpUrlConnection

The HttpUrlConnection class (found in java.net.HttpUrlConnection) is a light-weight HTTP client. The quick example we did at the beginning of this chapter uses this class. Be aware that prior to Froyo (2.1/2.2), there were a number of bugs associated with this class, so it is recommended that applications needing to support Froyo and Eclair should use the Apache HTTP Client for those platforms. Because HttpUrlCon nection is intended to be lightweight, it does not have all the nice convenience methods and wrapper classes that the Apache HTTP Client classes do. Connections are made directly and responses are read in directly from the InputStream that is established.

Example 15-14 shows a GET request. It doesn't use many of the trappings of HttpGet objects, but instead establishes the connection with an openConnection() method. The status code is read with a convenience method called getResponseCode(), and then the input is read directly.

Example 15-14. GET with HttpUrlConnection

```
URL url = new URL("http://www.someserver.com/");
HttpURLConnection urlConnection = (HttpURLConnection) url.openConnection();
try {
        int statusCode = urlConnection.getResponseCode();
        System.out.println("Response Code: "+statusCode);

        InputStream in =
                new BufferedInputStream(urlConnection.getInputStream());
        // do something with the InputStream
} finally {
```

```
        urlConnection.disconnect();
}
```

In the case of a POST, once again, using HttpUrlConnection is pretty simple. You can send a list of key-value pairs by stringing together the list and writing it out directly to the OutputStream. A method called setDoOutput() must be set to true. Request parameters such as Content-Type are specified directly as well.

Because we know the length of the data in this example (we find it through a data.get Bytes().length() call), we provide it to the connection using setFixedLengthStrea mingMode(). When the data size is unknown—such as when streaming data—and the server is HTTP/1.1-compliant, don't use setFixedLengthStreamingMode(). Instead, set a chunk size through setChunkedStreamingMode(0), as Example 15-15 illustrates.

Example 15-15. POST key-value list using HttpUrlConnection

```
String data = "key1=" + URLEncoder.encode("value1","UTF-8")+
              "&key2=" + URLEncoder.encode("value2","UTF-8");

URL url = new URL("http://www.someserver.com/");
HttpURLConnection urlConnection = (HttpURLConnection) url.openConnection();
try {
        urlConnection.setDoOutput(true);
        urlConnection.setRequestMethod("POST");
        urlConnection.setFixedLengthStreamingMode(data.getBytes().length()); i
        urlConnection.setRequestProperty("Content-Type",
                    "application/x-www-form-urlencoded");

        OutputStream out =
                new BufferedOutputStream(urlConnection.getOutputStream());
        out.print(data);
        out.flush();
        out.close();

        InputStream in =
                new BufferedInputStream(urlConnection.getInputStream());
        // do something with the InputStream
} finally {
        urlConnection.disconnect();
}
```

Unfortunately, the side effect of this simplicity is that when it comes to a slightly more complex thing such as posting a file as a multipart file stream, things get a tad complicated, as shown in Example 15-16. Essentially, you must provide some HTTP wrapper information manually as strings.

Example 15-16. POST file using HttpUrlConnection

```
String attachmentName = "somefile";
String attachmentFileName = "somefile.bmp";
```

```
String crlf = "\r\n";
String twoHyphens = "--";
File file = new File(attachmentFileName);

URL url = new URL("http://www.someserver.com/");
HttpURLConnection urlConnection = (HttpURLConnection) url.openConnection();
try {
        urlConnection.setDoOutput(true);
        urlConnection.setFixedLengthStreamingMode(file.getBytes().length()); i
        urlConnection.setRequestProperty("Content-Type",
                        "multipart/form-data;boundary=" + boundary");

        DataOutputStream request =
                new DataOutputStream(httpUrlConnection.getOutputStream());

        request.writeBytes(twoHyphens + boundary + crlf);
        request.writeBytes("Content-Disposition: form-data; name=\"" +
                                attachmentName + "\";filename=\"" +
                                attachmentFileName + "\"" + crlf);
        request.writeBytes(crlf);

        request.write(convertFileToBytes(file));   // must convert file to bytes

        request.writeBytes(crlf);
        request.writeBytes(twoHyphens + boundary + twoHyphens + crlf);

        request.flush();
        request.close();

        InputStream in = new BufferedInputStream(urlConnection.getInputStream());
        // do something with the InputStream
} finally {
        urlConnection.disconnect();
}
```

However, you could dispense with posting as a multipart file and stream the data directly, as shown in Example 15-17.

Example 15-17. POST file as stream using HttpUrlConnection

```
File file = new File("somefile.txt");

URL url = new URL("http://www.someserver.com/");
HttpURLConnection urlConnection = (HttpURLConnection) url.openConnection();
try {
        urlConnection.setDoOutput(true);
        urlConnection.setRequestMethod("POST");
        setChunkedStreamingMode(0);

        OutputStream out = urlConnection.getOutputStream();
        out.print(convertFileToBytes(file));   // must convert file to bytes
        out.flush();
```

```
        out.close();

        InputStream in =
                new BufferedInputStream(urlConnection.getInputStream());
        // do something with the InputStream
} finally {
        urlConnection.disconnect();
}
```

Networking in the Background using AsyncTask and AsyncTaskLoader

As discussed earlier, the various ways of doing an HTTP request/response transmission all use synchronized communication, and because this introduces delays, it cannot be done in the the UI thread. Therefore, you should run these methods in a background thread class such as `AsyncTask`. Because `AsyncTask` was covered previously in "AsyncTask" on page 116, we'll just provide an example using `HttpUrlConnection` in an `AsyncTask` (see Example 15-18).

Example 15-18. AsyncTask with HttpUrlConnection

```
private class GetConnectionTask extends AsyncTask<URL, Void, String>{
        @Override
        protected String doInBackground(URL... urls) {
                HttpURLConnection aHttpURLConnection =
                        (HttpURLConnection) url[0].openConnection();

                InputStream in = aHttpURLConnection.getInputStream();

                return convertInputStreamToString(in);
        }

        @Override
        protected void onPostExecute(String result) {
                // do something with the result
        }
}

// To use
GetConnectionTask task = new GetConnectionTask();
task.execute(new URL("http://www.someserver.com/"));
```

Summary

In this chapter, we took a brief step back to cover a fundamental piece of mobile development: HTTP network communication. The intent was to provide more detail such that you may be able to communicate with the larger world via the Internet and provide more capability with your application to your user.

Interaction and Animation: Live Wallpaper and Handlers

This chapter covers the animated LiveWallpaper API that allows developers to create interactive wallpaper that users may choose to run as part of their home page. We also cover handlers, an essential part of the Android thread system that enhances interactivity.

Live Wallpaper

Android 2.1 (API Level 7) introduced live wallpaper. A live wallpaper is a wallpaper (a background set on the home screen) that may be animated and enabled for interaction. It has access to the other services and APIs as normal Android applications: network, GPS, etc. The primary class to use when creating a live wallpaper is the `LiveWallpaper` `Service` (located in `android.service.wallpaper.WallpaperService`).

For an example, we will put together a live wallpaper that is touch-enabled and utilizes the Internet (via the Yamba Manager) in order to place some text at the point of touch. To do this, you must add a service definition within the main Manifest (see Example 16-1). Next, create a Service Resource (see Example 16-2).

Example 16-1. LiveWallpaper Service Manifest entry

```
<manifest xmlns:android="http://schemas.android.com/apk/res/android"
    package="com.marakana.android.yamba">

    <application>

        <service
            android:label="@string/app_name"
            android:icon="@drawable/ic_launcher"
            android:name=".YambaWallpaper"
            android:permission="android.permission.BIND_WALLPAPER">
```

```
        <intent-filter>
          <action
            android:name="android.service.wallpaper.WallpaperService" />
        </intent-filter>
        <meta-data android:name="android.service.wallpaper"
            android:resource="@xml/wallpaper" />
      </service>

  </application>
</manifest>
```

Example 16-2. LiveWallpaper Service Resource XML

```
<?xml version="1.0" encoding="utf-8"?>

<!-- the wallpaper.xml file is located in res/xml -->

<wallpaper xmlns:android="http://schemas.android.com/apk/res/android" />
```

Once the service references are created, write the `WallpaperService` code (Example 16-3). The `WallpaperService` has within it a special method called `onCreateEngine()`. It is called at creation time and returns a `WallpaperService.Engine` object. That object is responsible for handling the wallpaper's life cycle as well as the graphical and interactive aspects of the wallpaper. We'll explain the lengthy code in the sections that follow.

To achieve a simple animation effect, we'll be drawing on a canvas. The wallpaper will display recent Yamba status updates, and we'll interact with the user by letting her choose where on the screen the updates should appear. This code gives you a flavor of what user interaction can look like on Android, but drawing and animation in general are beyond the scope of this book.

Example 16-3. LiveWallpaper service

```
package com.marakana.android.yamba;

import java.util.List;

import com.marakana.android.yamba.clientlib.YambaClient;

import android.graphics.Canvas;
import android.graphics.Color;
import android.graphics.Paint;
import android.os.Handler;
import android.service.wallpaper.WallpaperService;
import android.view.MotionEvent;
import android.view.SurfaceHolder;

public class YambaWallpaper extends WallpaperService {

    @Override
```

```java
public Engine onCreateEngine() {
    return
    new YambaWallpaperEngine(((YambaApplication) getApplication()).
        getYambaClient());
}

private class YambaWallpaperEngine extends Engine implements Runnable {

    private Handler handler = new Handler();
    private ContentThread contentThread = new ContentThread();
    private YambaClient yambaclient;

    private Paint paint;

    private String[] content = new String[20];
    private TextPoint[] textPoints = new TextPoint[20];
    private int current = -1;
    private boolean running = true;
    private float offset = 0;

    public YambaWallpaperEngine(YambaClient client) {
        yambaclient = client;

        paint = new Paint();
        paint.setColor(0xffffffff);
        paint.setAntiAlias(true);
        paint.setStrokeWidth(1);
        paint.setStrokeCap(Paint.Cap.SQUARE);
        paint.setStyle(Paint.Style.FILL);
        paint.setTextSize(40);
    }

    @Override
    public void onCreate(SurfaceHolder surfaceHolder) {
        super.onCreate(surfaceHolder);
        running = true;
        contentThread.start();

            // enable touch events
        setTouchEventsEnabled(true);
    }

    @Override
    public void onDestroy() {
        super.onDestroy();
        handler.removeCallbacks(this);

        running = false;

        synchronized(contentThread) {
            contentThread.interrupt();
        }
```

```
    }

    @Override
    public void onVisibilityChanged(boolean visible) {
        if (visible) {
            drawFrame();
        } else {
            handler.removeCallbacks(this);
        }
    }

    @Override
    public void onSurfaceChanged(SurfaceHolder holder, int format,
                                          int width, int height) {
        super.onSurfaceChanged(holder, format, width, height);
        drawFrame();
    }

    @Override
    public void onSurfaceCreated(SurfaceHolder holder) {
        super.onSurfaceCreated(holder);
    }

    @Override
    public void onSurfaceDestroyed(SurfaceHolder holder) {
        super.onSurfaceDestroyed(holder);
        handler.removeCallbacks(this);
    }

    @Override
    public void onOffsetsChanged(float xOffset, float yOffset,
            float xStep, float yStep, int xPixels, int yPixels) {
        offset = xPixels;

        drawFrame();
    }

    @Override
    public void onTouchEvent(MotionEvent event) {
        if (event.getAction() == MotionEvent.ACTION_DOWN) {
            current++;

            if(current >= textPoints.length) {
                current = 0;
            }

            String text = content[current];
            if(text != null) {
                textPoints[current] =
                        new TextPoint(text, event.getX() -
                                offset, event.getY());
            }
```

```
        }
        super.onTouchEvent(event);
    }

    @Override
    public void run() {
            drawFrame();
    }

    private void drawFrame() {
        final SurfaceHolder holder = getSurfaceHolder();

        Canvas c = null;
        try {
            c = holder.lockCanvas();
            if (c != null) {
                // draw text
                drawText(c);
            }
        } catch (Exception e) {
            e.printStackTrace();
        } finally {
            if (c != null) {
                    holder.unlockCanvasAndPost(c);
            }
        }

        // Reschedule the next redraw
        handler.removeCallbacks(this);

        if (isVisible()) {
            handler.postDelayed(this, 40); // 40 ms = 25 frames per second
        }
    }

    private boolean getContent() {
            List<YambaClient.Status> timeline = null;

            try {
                    timeline = yambaclient.getTimeline(20);

                    int i = -1;
                    content = new String[20];
                    if(timeline != null) {
                            for(YambaClient.Status status: timeline) {
                                    i++;
                                    content[i] = status.getMessage();
                            }
                    }
            } catch (Exception e) {}
            return timeline != null && !timeline.isEmpty();
    }
```

```
private void drawText(Canvas c) {
        c.drawColor(Color.BLACK);

        for(TextPoint textpoint: textPoints) {
                if(textpoint != null) {
                        c.drawText(textpoint.text,
                                textpoint.x + offset,
                                textpoint.y, paint);
                }
        }

}

private class TextPoint {
        public String text;
        public float x;
        public float y;

        public TextPoint(String t, float xp, float yp) {
                text = t;
                x = xp;
                y = yp;
        }
}

private class ContentThread extends Thread {
        public void run() {
                while(running) {
                        try {
                                boolean hascontent = getContent();
                                if(hascontent) {
                                        Thread.sleep(60000);  // 1 min
                                } else {
                                        Thread.sleep(2000);  // 2 s
                                }
                        } catch (InterruptedException ie) {
                                return;
                        } catch (Exception e) {}
                }
        }
    }
  }
}
```

This long sample of code defines four classes: YambaWallpaper, YambaWallpaperEn
gine, TextPoint, and ContentThread. YambaWallpaper is the primary class, extending
the base class WallpaperService. Within YambaWallpaper, we create the YambaWallpa
perEngine private class and create it via the onCreateEngine() method.

YambaWallpaperEngine itself contains the other two classes. TextPoint is a data structure containing a string (the text) and two coordinates (x and y) that represent screen locations. ContentThread is a special thread that loops continuously. At regular intervals it invokes the getContent() method, which attempts to get YambaClient.Status objects via the getTimeline() method in the YambaClient. These objects contain text statuses that have been sent to the Yamba API Web Service. If content comes back, we store the text of these statuses and then wait a minute before checking for content again. If there is no content during this call, we wait two seconds and see again whether we can get content.

YambaWallpaperEngine is the most important class within this example, however. As a class that extends android.service.wallpaper.WallpaperService.Engine, it is the piece of code that handles the life cycle and drawing that occurs in the wallpaper. When the engine is created, the runtime calls its onCreate() method, where we start the ContentThread. After the engine is created, the next major event is making the wallpaper visible, which takes place in onVisibilityChanged(). At this point, the first call to drawFrame() is made. This method hooks into the underlying Canvas (the paintable screen), and we proceed to draw upon it through the drawText() method. draw Frame() is also called after someone starts an app and then puts it into the background, so that the wallpaper becomes visible again. A change in visibility is reported to us by the runtime, which calls onVisibilityChanged(). Should there be a change to the SurfaceHolder (the display surface), then surfaceChanged() is called. When that occurs we also do a drawFrame() call.

drawText() draws recently received updates on the screen. The color and style of the text are taken from the Paint object generated in the constructor for YambaWallpaper Engine.

The text that is to be drawn is derived from the TextPoint objects in the textPoints array. These objects are generated in the onTouchEvent() method that is triggered when the user touches the wallpaper's screen surface. When the user touches the screen, we capture the touch event's x and y positions as well as select one of the list of text statuses that was received via the getContent() method. Each element of the textPoints array thus contains an x and y position where text will be drawn, along with a string to hold a recent status update. We step through the array, drawing a new text each time the user touches the wallpaper.

In the drawFrame() method we get the surface holder using getSurfaceHolder(), lock the canvas using holder.lockCanvas(), draw on the canvas, and then unlock it and push the changes through holder.unlockCanvasAndPost(). By repeating this over and over every 40 ms, we generate an ongoing animation at 25 fps (frames per second). One thing to note is that when the SurfaceHolder is destroyed, onSurfaceDestroyed() is called, and we break the animation cycle.

Handler

Example 16-3 uses a `Handler` for interaction. Every `Handler` is associated with a `Loop er` and through that `Looper` a special kind of thread and that thread's message queue. This special kind of thread extends the generic Java thread. In Android, this special thread may have an associated `MessageQueue` that is established via a call to the `Loop er` class. This establishes a queue within which `Message` objects are placed and may be retrieved and handled by a `Handler`. Through this mechanism, other threads may communicate with the thread that has the message queue. The Main UI thread is an example of such a thread with a message queue.

You can associate handlers explicitly when the `Handler` object is instantiated via a direct reference to the constructor in the `Handler`:

```
new Handler(Looper looper)
```

If you pass no argument, the `Handler` is associated with the thread within which it is created. In the wallpaper example, the `Handler` is instantiated at the time the `Yamba WallpaperEngine` is instantiated via `onCreateEngine()`, so the `Handler` is associated with the Main UI thread. This `Handler` handles the Main UI's message queue and thus is able to interact with the main UI.

One way to use a `Handler` is to post a `Runnable` to it (see Example 16-4). This is what the wallpaper example did, but it invoked the `postDelayed()` method, which adds a specified delay in milliseconds prior to running.

Example 16-4. Handler post Runnable

```
Handler handler = new Handler();

handler.post(new Runnable() {
                @Override
                public void run() {
                    // do something
                }
            });
```

To dig further into how exactly the `YambaWallpaperEngine` uses its `Handler`, take a look at Example 16-5. This example is a simpler version of what the `YambaWallpaperEn gine` is doing. The `Handler` is created with the name `handler`, a `Runnable` object named `runner` is created, and this `Runnable` object is passed to the handler's `postDelayed` method. That adds the runner to the message queue.

Forty milliseconds later, the `Runnable`'s run method is called. It prints the word "run" and then removes the callback method reference in the handler. This `removeCall backs` method ensures that the `Runnable` is cleared out of the message queue. (In the live wallpaper example, for instance, we call `removeCallbacks` in the `onDestroy()`

---

method in case the service is being destroyed, and thus stop the execution of that Runnable from occurring.) Then we generate a random true or false value, and if it is true, we add the runner back into the message queue to start all over again. Otherwise, the run method ends and the program completes.

Example 16-5. Handler postdelayed and remove

```
final Handler handler = new Handler();

final Runnable runner = new Runnable() {
                    @Override
                    public void run() {
                          System.out.println("run"):

                          handler.removeCallbacks(runner);

                          // random true or false
                          if((new Random()).nextBoolean()) {
                                    // if true, do a post delay of
                                    //  40 ms and run the run again
                                    handler.postDelayed(runner, 40);
                          }
                    }
             };

handler.postDelayed(runner, 40);
```

A second way to use a Handler is to post a Message object to the message queue (Example 16-6). The message is handled either by a specified Handler.Callback or by the Handler's own handleMessage().

Example 16-6. Handler handleMessage

```
Handler handler = new Handler() {
       @Override
       public void handleMessage(Message msg) {
             // do something with msg
       }
};
```

A Message object has a variety of fields and methods that you can use to pass message content to the callback object (see Example 16-7).

Example 16-7. Handler handleMessage detailed example

```
final Handler handler = new Handler() {
       @Override
       public void handleMessage(Message msg) {
             if(msg.what == 1) {
                    System.out.println("Msg: "+((String) msg.obj));
                    doSomethingOnMainUIThread(msg.obj);
             }
```

```
        }
};

Thread someThread = new Thread() {
        @Override
        public void run() {
                Messege msg = Message.obtain();
                msg.what = 1;
                msg.obj = new String("Some String");
                handler.sendMessage(msg);
        }
};
```

Although the `Message` constructor is public and thus could be directly instantiated, you should call either `Message.obtain()` or `Handler.obtainMessage()` (or one of its derivatives) to get a `Message` object from a global pool of recycled objects to reduce resource usage.

Summary

In this final chapter we introduced some of the concepts used for doing animation and creating effects in the background. Handlers allow delayed events, as well as those that occur at regular intervals. A wide range of techniques can be used to spruce up the user's main screen through LiveWallpaper.

Index

We'd like to hear your suggestions for improving our indexes. Send email to index@oreilly.com.

instances, 13
int data types, 13
intent broadcasts, 223
intent filters, 225
intent services, 83
intents, 68, 82
    broadcasting, 227–231
    explicit, 68
    implicit, 68
IntentService class, 166
internet permissions, 113

J

Java, 9–29
    Android and, 37–39
    arrays, 15
    collections, 27
    comments, 12
    complex example sample code, 22–26
    constructors, 14
    control flow statements, 16–18
    data types, 13
    error handling, 19–22
    exception handling, 19–22
    generics, 28
    inheritance, 26
    interfaces, 26
    modifiers, 14
    object data types, 13
    objects, initializing, 106
    operators, 16
    primitive data types, 13
    program basics, 9–12
    threading in, 28
    user event handlers, 107
    XML, inflating to, 104–106
Java Development Kit, 43–45
    on Linux, 44
    on Mac, 44
    on Windows, 44
Java Development Kit (JDK), 9
Java Enterprise Edition, 39, 44
Java Micro Edition, 39
Java Mobile Edition (JavaME), 44
Java Runtime Environment (JRE), 44
Java Standard Edition, 39
Java Virtual Machine (VM), 36
JavaSE, 44

K

keys, 142
Kindle Fire, 5

L

layout folder, 55
layout gravity (properties), 102
layout height (properties), 101
layout weight (properties), 102
layout width (properties), 101
layouts, 88–92
    common properties for, 101
    FrameLayout, 91
    LinearLayout, 89
    RelativeLayout, 92
    strings resources, 103
    TableLayout, 90
Lesser General Public License (LGPL), 35
libraries, native, 34
life cycle methods, 164
    onCreate(), 164
    onDestroy(), 165
    onStartCommand(), 165
LinearLayout, 89
Linux features, 33
Linux kernel, 31
Linux portabilities, 32
Linux securities, 32
Linux vs. Android, 33
listeners, 119–124
lists, 83
live wallpaper, 85, 253–259
LogCat mechanism, 108
logging, 108
    LogCat mechanism, 108
long data types, 13
loops, 17

M

Macintosh, installing Java on, 44
Main Activity, creating, 149
mangers, 39
manifest file, 40, 51–54
    adding activities to, 147
    adding content providers to, 199
    adding services to, 167
    internet permissions and, 113

registering broadcast receivers in, 225
widgets and, 238
media servers, 36
menu events, handling, 154
menu resource, 150–152
menu system, 82
messages, 108
methods, 13
modifier (method), 14
Motorola, 7
multithreading, 81, 115

N

name (method), 14
native layer, 33–36
native daemons, 35
native libraries, 34
native tools, 36
native libraries, 41
network receivers, 84
networking, 81, 85, 242–244
AsyncTask and, 251
AsyncTaskLoader and, 251
HTTP API, 244
nonaccess modifiers, 14, 15
notifications, receiving, 230

O

OAuth, 76
object data types, 13
object-oriented programming, 13
onCreate() (method), 164
onCreateView() (method), 136
onDestroy() (method), 165
onDestroyView() (method), 136
onPause() (method), 136
onResume() (method), 136
onStartCommand() (method), 165
Open Handset Alliance, 5
OpenGL, 35
OpenSSL, 35
operators, in Java, 16
OPTIONS (method), 243
original equipment manufacturers (OEMs), 7
overloading, 27

P

package names, 94
partitions, 157–161
SDCard partition, 158
system partition, 158
user data partition, 160
paused state, 66
permissions, 84
POST (method), 243
preference activity, 82
preferences, 141–148
adding to manifest file, 147
implementing, 84
resources for, 142–145
shared, 155–157
primitive data types, 13
private modifiers, 15
programmatic user interface, 88
project design, 80
project, anatomy of
drawable resources, 56–58
layout folder, 55
manifest file, 51–54
string resources, 54
projects, naming, 93
properties, 13
protected modifiers, 15
public modifiers, 15
publishers, 223
PUT (method), 243

Q

query() (operation), 177

R

R.java file, 56
radio interface layer daemon, 35
refactoring code, 80
relative time, 211
RelativeLayout, 92
res folder, 58
resources, 40
alternative, 124–127
Android system, 152
menu, 150–152
strings, 103
resources.ap, 59

return statements, 18
return type (method), 14
running state, 66

S

Samsung, 7
SDCard partition, 158
SDK version, 94
service manager, 35
services, 68, 82, 163–172
    adding to manifest file, 167
    connecting to databases with, 179–181
    creating Java classes for, 164–166
    intent, 83
    IntentService class, 166
    menu handling of, 168
    pulling data with, 169–172
    starting/stopping, 168
    testing, 169
shared preferences, 155–157
short data types, 13
signatures, 41
single thread execution, 114
SQL support, 83
SQLite, 35, 176
sqlite3, 184
SQLiteOpenHelper class, 176
stack, 31–42
    anatomy of, 31
    application framework, 39
    applications, 40–42
    Dalvik, 36–39
    HAL, 34
    Linux kernel, 31
    native layer, 33–36
starting state, 65
states (of activities), 64–67
static initialization, 137
static modifiers, 15
statically typing, 13
stopped state, 66
string resources, 54
subscribers, 223
summaries, 142
system partition, 158

T

TableLayout, 90

TAG, 108
target SDK, 94
text (properties), 102
threading, 114–119
    AsyncTask class and, 116–119
    in Java, 28
    multi-, 115
    single thread, 114
throwing exceptions, 21
timeline receivers, 84
titles, 142, 151
tools, native, 36
TRACE (method), 243
Twitter
    API, history of, 111
    Yamba vs., 76

U

update() (operation), 178
user data partition, 160
user interface, 81, 87–127
    alternative resources for, 124–127
    creating, 87
    declarative, 87
    designing in Eclipse, 97–102
    events, handling, 107
    implementing, 104–108
    layouts, 88–92
    listeners, 119–124
    programmatic, 88
    views, 88–92

V

variables, 13
views, 88–92

W

web protocols, 241–251
    Apache HTTP client, 245–247
    AsyncTask and, 251
    AsyncTaskLoader and, 251
    HTTP API, 244
    HttpUrlConnection class, 248–250
    networking, 242–244
web services, 85
Webkit, 35
while loop, 17

About the Authors

Marko Gargenta is the director of Twitter University, where he manages the training of Twitter Engineers in Android and other open source technologies. Previously he was cofounder of Marakana (acquired by Twitter), a firm that trained thousands of Android developers at Intel, Cisco, Qualcomm, Motorola, the Department of Defense, and other institutions. Marko is also the creator of Android Bootcamp course and cofounder of San Francisco Android Users' Group.

Masumi Nakamura, VP of Engineering at Placester, Inc., has spent over 15 years in software doing everything from mobile development and scaling large backend systems, to running a data science team at Paypal. He also spends a lot of his time advising and working closely with a variety of startup companies.

Colophon

The animal on the cover of *Learning Android, Second Edition* is a Little Owl.

The Little Owl is part of the taxonomic family Strigdae, which is informally known as "typical owl" or "true owl" (the other taxonomic family includes barn owls). True to its name, the Little Owl is small, measuring between 23 and 27.5 centimeters in length. It is native to the warmer areas of east Asia (particularly Korea), Europe, and North Africa and has been introduced and naturalized in Great Britain and the South Island of New Zealand.

The Little Owl is characterized by long legs and a round head with yellow eyes and white eyebrows; the eyebrows are said to give the owl a serious expression. The most widespread species, *Athene noctua*, is white and speckled brown on top and white-and-brown streaked on bottom. A species commonly found in the Middle East, *A. n. lilith*, or the Syrian Little Owl, is a pale grayish-brown.

The sedentary Little Owl typically makes its home in open country, such as parkland and farmland. It preys on amphibians, earthworms, insects, and even smaller mammals and birds; despite its diminutive stature, the Little Owl is able to attack many game birds. Unlike many of its true owl family members, the Little Owl is diurnal, or active during the day, during which it often perches openly. Depending on the habitat, the Little Owl builds nests in cliffs, rocks, holes in trees, river banks, and buildings. Little Owls that live in areas with human activity tend to get used to people and may perch in full view when humans are present.

The cover image is from Cassell's *Natural History*. The cover fonts are URW Typewriter and Guardian Sans. The text font is Adobe Minion Pro; the heading font is Adobe Myriad Condensed; and the code font is Dalton Maag's Ubuntu Mono.

Get even more for your money.

Join the O'Reilly Community, and register the O'Reilly books you own. It's free, and you'll get:

- $4.99 ebook upgrade offer
- 40% upgrade offer on O'Reilly print books
- Membership discounts on books and events
- Free lifetime updates to ebooks and videos
- Multiple ebook formats, DRM FREE
- Participation in the O'Reilly community
- Newsletters
- Account management
- 100% Satisfaction Guarantee

Signing up is easy:

1. **Go to: oreilly.com/go/register**
2. **Create an O'Reilly login.**
3. **Provide your address.**
4. **Register your books.**

Note: English-language books only

To order books online:
oreilly.com/store

For questions about products or an order:
orders@oreilly.com

To sign up to get topic-specific email announcements and/or news about upcoming books, conferences, special offers, and new technologies:
elists@oreilly.com

For technical questions about book content:
booktech@oreilly.com

To submit new book proposals to our editors:
proposals@oreilly.com

O'Reilly books are available in multiple DRM-free ebook formats. For more information:
oreilly.com/ebooks

Spreading the knowledge of innovators   oreilly.com

CPSIA information can be obtained at www.ICGtesting.com
Printed in the USA
BVOW10s1804290114

343310BV00003B/3/P

9 781449 319236